Village Water Supply in the Decade

Community participation—how does it work?

(Photo: L. H. Robertson)

Village Water Supply in the Decade

Lessons from Field Experience

Colin Glennie
UNICEF, Kathmandu, Nepal

A Wiley-Interscience Publication

JOHN WILEY & SONS
Chichester · New York · Brisbane · Toronto · Singapore

Copyright © 1983 by John Wiley & Sons Ltd.

All rights reserved.

No part of this book may be reproduced by any means, nor transmitted, nor translated into a machine language without the written permission of the publisher.

Library of Congress Cataloguing in Publication Data

Glennie, Colin, 1946–
   Village water supply in the decade.

   'A Wiley-Interscience publication.'
   Bibliography: p.
   Includes index.
   1. Underdeveloped areas—Water-supply. 2. Water-supply, Rural—Malawi. I. Title.
TD927.G54 1983   363.6'1'091734   82–23749

ISBN 0 471 10525 2

British Library Cataloguing in Publication Data:

Glennie, Colin
   Village water supply in the decade.
   1. Underdeveloped areas—Water supply
   2. Water supply, Rural
   I. Title
   628'.72'091724   TD481

ISBN 0 471 10525 2

Typeset by Photo-Graphics, Honiton, Devon.
Printed by Page Bros. (Norwich) Ltd.

Contents

Foreword		ix
Preface		xi

1 INTRODUCTION — 1
1.1 The Village Perspective — 1
1.2 The Decade Perspective — 2
1.3 Background to Malawi — 4

2 HISTORICAL DEVELOPMENT — 11
2.1 Introduction — 11
2.2 Pilot Phase — 11
2.3 Consolidation Phase — 19
2.4 Expansion of the Programme — 21
2.5 The Role of Donor Agencies — 22
2.6 Conclusion — 22

3 PROGRAMME ORGANIZATION AND MANAGEMENT — 25
3.1 Introduction — 25
3.2 Objectives — 25
3.3 Organization of the Programme — 25
3.4 Programme Planning — 30
3.5 Project Selection Criteria — 33
3.6 Project Preparation — 35
3.7 Supplies — 35

4 FIELD LEVEL ORGANIZATION AND MANAGEMENT — 37
4.1 Introduction — 37
4.2 Project Organization — 38
4.3 Project Planning — 39
4.4 Community Participation — 42
4.5 Management of Self-help Labour — 45
4.6 Management of Project Staff — 56
4.7 Tank Construction — 58
4.8 Transport — 62
4.9 Project Stores Organization — 64

5	FIELD STAFF	65
	5.1 Introduction	65
	5.2 Policy	65
	5.3 Selection Criteria for Field Technicians	66
	5.4 Recruitment Procedure	67
	5.5 Selection/Training Course	68
	5.6 On-the-job Training	71
	5.7 Field Technicians' Career Structure	72
	5.8 Motivation of Staff	73
	5.9 Supervisors	75
	5.10 Engineers	76
	5.11 Conclusion	78
6	TECHNICAL ASPECTS	79
	6.1 Introduction	79
	6.2 Design	79
	6.3 Construction Techniques	82
	6.4 Technical Description of a Major Project	83
	6.5 Asbestos Cement Pipes	91
	6.6 PVC Pipes	92
	6.7 Steel Pipes	94
	6.8 Polyethylene Pipes	94
	6.9 Technical Developments	94
	6.10 Summary	96
7	MAINTENANCE	97
	7.1 Introduction	97
	7.2 Factors Affecting Maintenance	97
	7.3 Maintenance Policy in Malawi	99
	7.4 An Appropriate Maintenance Policy	101
	7.5 Maintenance and Repair Load	105
	7.6 Monitoring and Evaluation	109
8	BENEFITS	111
	8.1 Introduction	111
	8.2 Benefits of the Malawi Programme	111
	8.3 Improving Health Benefits	113
9	SANITATION	115
	9.1 Introduction	115
	9.2 Strategy	115
	9.3 Stimulating Demand	116
	9.4 Adoption Process	120
	9.5 Technical Factors	120
	9.6 Conclusion	122

10 IMPLICATIONS FOR THE DECADE 123
10.1 Introduction 123
10.2 Field Management 123
10.3 The Four Phases of Programme Development 124
10.4 Conclusion 129

APPENDICES

1 Rural Piped Water Projects in Malawi at January 1979 131
2 Job Descriptions of the Malawi Programme 133
3 Design Procedure for Gravity Piped Water Systems in Malawi 137
4 Example of Use of Topographical Maps in Malawi 145
5 Example of Use of Aerial Photographs in Malawi 147

BIBLIOGRAPHY 149

Index 151

Foreword

These are the early years in the International Drinking Water Supply and Sanitation Decade. As yet, it is too soon to predict whether the Decade will make significant headway towards its goal of clean water and adequate sanitation for all by 1990. External pressures from the UN and aid agencies are being felt, developing countries are planning expanded programmes, raising targets, and giving increased priority and funding for this sector. There is some danger, however, that national water and sanitation programmes will not be able to grow rapidly enough to cope with the expanded responsibilities and influx of funds.

Emphasis is now given to rural development and meeting the needs of low-income populations. Water supply has become viewed as a basic need, if not a right. Local experience on which future programmes must be built is largely confined to urban projects; unfortunately rural programmes which use primarily an urban or engineering approach have a marked propensity for failure. Few countries have developed successful self-help strategies applicable to rural areas, most are planner-designed and not "village-based". There are very few successful examples from which expanding rural water programmes can learn and emulate.

Colin Glennie describes the Malawi gravity water programme in depth. Having spent several years as a self-help project manager he is well qualified to do so. The detail he provides is welcome. There is already quite enough literature of a general nature giving overall planning guidelines but of little help to the engineer battling from day to day with practical problems in the field. I first met Colin on-site in Mulanje. Having visited several self-help schemes in the past I was more than a little sceptical of this one which, it was claimed, involved tens of thousands of people. We set off early to inspect the self-help construction in progress, but our Land Rover was held up by a river in flood and by the time we reached the site in the late morning we found the villagers were already departing, having completed their task for the day. They had risen before dawn, walked several kilometres to the site, dug several hundred metres of trench and were now returning to the fields to tend their crops. My scepticism soon dissipated; this proved to be the most successful self-help water supply scheme in Africa.

Typically, the self-help programme relies heavily on the personal enthusiasm and day-to-day supervision of its creator. Consequently, it seldom grows beyond that individual's personal sphere of influence and managerial capability. Lindesay Robertson conceptualized and designed the Malawi programme to avoid this pitfall and guided it through its early stages of development starting with a pilot project serving 3,000 people in Chingale in 1969 and eventually growing to cover over 600,000 throughout Malawi today. It has been said that the programme benefited from a set of specifically local factors which were conducive to its development, hence making it unique to Malawi and not replicable elsewhere. This is true to some extent as the Malawi model was not designed for export to other countries. There are, however, several prerequisites to success which are also applicable to other countries identified early in the programme and which are noteworthy here. These include:

(1) reliance on local initiative and involvement of the community throughout the planning process; use of a local committee structure which places responsibility for the project on those who are to benefit most;
(2) the project's integration into local government institutions;
(3) equal attention given to socio-political and technical detail;
(4) the creation of cohesive motivated teams of field assistants chosen from the area and given focused in-service training;
(5) strong managerial and logistic support in the field;
(6) installation of well-designed maintenance schemes which rely on the community and its assumption of long-term responsibility for the water supply installation; and finally,
(7) strong central government commitment to the welfare of its rural people.

These are several elements of the Malawi example which can be transferred and adapted to local conditions in other countries. Many more are given in the text. Certainly two of the more fundamental prerequisites for success in self-help projects are: (1) that they be given adequate time to grow from pilot to full-scale schemes (often this is far more time than allowed by executing agencies attempting to meet tight construction schedules); and (2) that they must be village-based starting from the initial request for assistance, through construction to long-term operation and maintenance by the community itself.

Yes, there is a great deal we can learn from the Malawi experience to the benefit of our own rural communities.

MICHAEL G. MCGARRY
Cowater International
Ottawa

Preface

Development is a human story. Its characters are the millions all over the globe who are engaged in development programmes. Development workers hardly have time to read a technical book, let alone write one. Their story is rarely told. The tales of frustration and mistakes, hard work, patience, and personal dedication tend to pass unnoticed in the literature of development.

I have had the opportunity to write this book, but it is not my story alone. It is the story of a remarkable team of engineers, supervisors, technicians, and artisans who were brought together and motivated by the vision and leadership of the man who started the programme in 1968. Lindesay Robertson is one of the many characters in development who seek neither credit nor publicity, yet without whom no development programme can ever succeed.

I was fortunate to be a member of the team for most of the period covered by this book. This is, therefore, a first-hand account of the development of a programme. I have deliberately not updated the material on the programme to include developments since I left Malawi, as I believe this should remain a first-hand account covering the first 10 years.

In addition to giving a detailed account of one particular programme, I have used the experience to draw out certain fundamental lessons for the development of rural water supply programmes which I firmly believe are of major relevance to similar programmes all over the developing world, particularly in this International Drinking Water Supply and Sanitation Decade.

I am grateful to the British Overseas Development Administration for permission to publish this book, which is based on material written during a course of study undertaken with their support. I thank Lindesay Robertson and Henk van Schaik for their suggestions and support, and for some of the material in this book. I also thank Dr Richard Feachem for his constructive criticism of the original manuscript, and Dr Michael McGarry for generously agreeing to write a Foreword.

The opinions expressed in this book are my own and do not necessarily reflect the views of the Malawi Government or any of its employees.

January 1983

COLIN GLENNIE
Kathmandu
Nepal

CHAPTER 1

Introduction

1.1 The Village Perspective

Most developing countries have rural water supply programmes. Some of these programmes are able to cope with the problems they encounter; others are overwhelmed by them, but are compelled by the political demand for water to continue operating. What makes some programmes more successful than others? This book describes some of the more important features for success.

In order to understand the type of problems that frequently occur all over the world, let us first consider a typical project in a village in a developing country where the rural water supply programme is not running so well.

Before the project, the women of the village spend hours every day drawing water. Due to increasing cultivation and erosion there is less water in the source each dry season; one day the source will dry up. Water collection is the most time-consuming and exhausting of all domestic chores. The water itself is polluted and, because of the effort involved, water is precious and is used sparingly, making adequate personal and home hygiene virtually impossible; so the burden of disease is added to the burden of water collection.

A local politician, anxious to demonstrate his ability to bring benefits to the people before the forthcoming election, uses his influence with other politicians and government officials and manages to get the village included in the rural water supply programme. The villagers themselves know nothing of this request until several months later, when news spreads around the village that some government people have come and taken measurements. Most of the villagers were away in the fields and only a few people saw or spoke to them. The visitors only stayed a few hours and did not say very much, so nobody knows what will happen next. Six months later the news spreads again that someone from the government has come and told one of the leaders that the project will start soon. The people do not believe this, but three weeks later someone who calls himself a "technician" comes with some pipes and other materials. He immediately asks why they have not yet started digging the trench, as they were told to by the supervisor who had visited three weeks earlier. The villagers know nothing about this, and are surprised that they are expected to provide their labour without pay as they know that people were paid to dig the trench for the water supply at the district centre. There is a heated discussion in the village; some people blame the leader to whom the supervisor spoke for not passing on the instructions to start digging; but the leader denies that he was told this; some say that they will refuse to dig the trench unless they are paid; others say they are too busy in the fields to dig a trench anyway. But the local politician manages to persuade the majority to cooperate and all the villagers eventually agree to provide their labour, except for one group which lives near the source. The villagers go ahead with digging the trench while the technician goes off to work in another village. When he returns several weeks later he tells them the trench is not deep enough and refuses to lay any pipe. This time the villagers become angry with him. He is young and seems to consider himself superior to the villagers; he never told them how to dig the trench and they suspect he is trying to show his power over them. In fact, the technician feels uncomfortable in the village; the people seem very uncooperative and troublesome and he is rather frightened by them; he has never experienced this sort of situation before and his training, which he has just completed, was mainly technical; his supervisor never visits the project and he feels very isolated and without support. To please the villagers, he agrees to lay the pipe even though the trench is not deep enough. But when there is not enough pipe to finish the line the villagers again become angry and accuse him of selling the pipe. This is not true, but the technician hurriedly returns to the district centre to collect more pipe. His supervisor complains that the project is behind schedule, and tells him to finish it off quickly as he is needed to start two new projects immediately. The technician returns to the village with more pipe and the people are very happy when water at last reaches the village. He leaves some materials behind and tells them to construct proper aprons and drainage at all the tap sites and to finish backfilling the line. He then reports to his supervisor that the project is finished and is immediately sent to start the two new projects. The "completed" project is marked on a chart in the ministry headquarters.

Back in the village, the people consider that they have achieved their objective and feel no motivation to build proper tapstands or finish backfilling the line. Within a few months the rains come and the pipeline is damaged in several places where the trench is not deep enough. The villagers carry out makeshift repairs each time there is a break, and the water flows intermittently throughout the first wet season. The next dry season, however, a dispute arises when the group nearer the source, who did not participate in the project, say they want a tap of their own. When the rest of the village refuse, saying they did not help to construct the system, the first group cuts the pipe. Each time the villagers try to repair it, it is cut again. The matter is reported to the district authorities, but no action is taken. The next wet season there is more damage to the pipeline but this time there seems no point in repairing it as water is not flowing anyway.

Life returns to normal. The women of the village again draw water from their original source.

1.2 The Decade Perspective

Most people who have worked in rural water supply programmes will be familiar with the sort of problems described above. Similar problems continue to afflict many programmes today. There is a great danger that the expansion of activity generated by the International Drinking Water Supply and Sanitation Decade will lead to such problems being repeated on an increasingly wide scale. Most governments of developing countries have adopted ambitious targets to extend water supply and sanitation coverage to most, if not all, of their populations by 1990.

Often with the assistance of donor agencies, governments are dramatically expanding their programmes with substantially increased allocations of money, manpower, and materials.

But this approach displays a false understanding of the nature of a rural development programme. Money, manpower, and materials do *not* themselves *constitute* a programme; increasing any, or all, of these does not *constitute* growth. A rural development programme can be likened to a living organism. Money, manpower, and materials are the *nutrients* rather than the constituent parts. If it is healthy, the organism can grow by metabolizing the nutrients at a certain rate, which will increase with growth. If, on the other hand, the organism is unhealthy one of two things may happen. If it receives exactly the right nutrients at a rate at which it can absorb, it will recover and become healthy. If, however, it receives excessive amounts of the "standard" nutrients it will never recover and will eventually die, having become too weak to compete with other organisms for nutrients.

The project described above was technically simple and the people genuinely wanted water. The failure was not due to lack of money, materials or, at least at the village level, manpower. The problem was that the programme was unhealthy—the village had not been involved properly in the request or survey; the technician was the wrong type of person for the job; his training was inadequate; he was given no support or supervision in the field; he had no system or set of procedures to follow; there was no procedure for inspecting the completed system or for instituting maintenance.

Much of the resources mobilized for the Decade are likely to be wasted unless they are used to help the development of strong, healthy programmes enjoying the confidence of the communities, with realistic and proven development policies, operating within well-organized institutional frameworks, with cadres of well-trained and experienced staff following well-established procedures. To achieve this, one of the most important resources for the Decade must be the exchange of information and experience, so that programmes experiencing problems can utilize the experience from other countries, and so that newly developing programmes can avoid the mistakes of others.

Unfortunately, in the growing volume of literature on rural water supply in developing countries there are few, if any, detailed first-hand accounts of a specific programme showing how it developed and how it works, what problems it faces, and how they are overcome. It is as if all the literature concerning, say, the motor car was limited to comparisons and studies of different cars by motor correspondents without a single detailed specification of a particular car by its manufacturer. One of the principal purposes of this book is to fill this gap and to provide engineers, planners, managers, and students with at least one specific example.

The second, more important purpose of this book is to indicate certain fundamental principles and guidelines for the development of a rural water supply programme in any developing country. To deduce these from the experience of one programme may at first seem presumptuous. However, the validity of such a deduction can be understood if one considers that the principles on which, say, a motor car engine works can certainly be deduced by examining the engine of one particular car. It is not, of course, valid to deduce that all programmes must develop in exactly the same way; obviously there is a wide range of situations existing in developing countries. Nevertheless, the author is convinced that the fundamental issues brought out in this book concerning the nature of the development process, management at central and particularly at field levels, field manpower and staff training, community participation, and the maintenance of completed water systems are not specific to Malawi but are generally relevant to all developing countries.

Although the programme described here concerns the construction of gravity-fed piped water systems, the fundamental principles and guidelines are not exclusive to this particular technology and are generally applicable to programmes installing handpump tubewells, dug wells or more sophisticated piped systems. Indeed, they are applicable in general terms to all rural development programmes which are based on community participation.

1.3 Background to Malawi

While the fundamental principles are applicable to all developing countries, the conditions existing in each country will obviously affect the ease with which they can be applied. To give the reader an understanding of the conditions under which the programme developed in Malawi, a brief account is given here of the geographical, political, social, and administrative situation in the country as far as it is relevant to the discussion.

National profile

Malawi is a long, narrow, land-locked country in southeastern Africa, with a land area approximately three-quarters that of England and Wales. The country is divided into three geographically distinct regions. The Northern Region is mountainous with altitudes up to 2,400 metres, the Central Region is largely flat plateau of about 1,400 metres and the Southern Region consists of highlands and mountains rising to 3,000 metres from low-lying plains. The hot, wet season lasts from November to April and is followed by a relatively cool dry season. There are great variations in rainfall and temperature, even between neighbouring districts.

In 1977 the population of Malawi was 5.57 million, with an average annual growth rate of 2.9 per cent. The mean population density was 59.2 persons per square kilometre, varying from 250 to 15 in the most and least populated districts. These figures are higher than most other countries of Africa but much lower than most of Asia. Over 90 per cent of the population live in traditional rural villages.

Since its independence from Britain in 1964, Malawi has been characterized by an unusually strong central government under the leadership of the President, Dr H. Kamuzu Banda. There is a strong political discipline in the country which, whatever one's political persuasion, has resulted in a favourable climate for national development. The Civil Service is generally competent, efficient, disciplined, and incorrupt. Being a relatively small country, Malawi is not as plagued with excessive bureaucracy as many other countries of the developing world. Cooperation between ministries is comparatively good as all development activities under sectoral ministries are coordinated by a very strong Development Division in the Office of the President and Cabinet, which is responsible for the administration of the national development plan. Malawi has no mineral resources and so development expenditure has concentrated on national infrastructure (transport, communications, and power) and agriculture.

With a per capita gross national product (GNP) of US$170 (1977), Malawi is among the 25 poorest nations of the world. Ninety per cent of the development budget comes from foreign loans and grants. Because of this, and the significant exports of tobacco and tea, Malawi is less afflicted than some countries with the problems of foreign exchange. Agriculture occupies 85 per cent of the labour force and accounts for 50 per cent of the GNP. The manufacturing sector is very small but it is expanding and includes a small PVC pipe extrusion plant. Communications within the country are relatively good, with an extensive network of fairly well-maintained roads. Village tracks are generally passable by light vehicles in the dry season. Malawi is heavily dependent on its rail links through Mozambique to the Indian Ocean ports of Beira and Nacala.

In 1966 adult literacy was 22 per cent; in 1976 primary school enrolment was about 50 per cent, with about 2 per cent of pupils proceeding to secondary school. The Malawi Polytechnic provides technical and business diploma courses and the University provides a general degree course. Because of the acute lack of trained professional and technical manpower, expatriates still play an important role in development work.

Malaria, respiratory, diarrhoeal and other gastro-intestinal diseases are the leading causes of sickness among rural people. Life expectancy was 42.5 years (1970), crude death rate 28.2 (1978) and infant mortality 163.8 (1978) per thousand. 29.6 per cent of children under five years were below 80 per cent Harvard Nutritional Standard.

Rural areas

At district level, the senior government official is the District Commissioner, who is responsible for upholding both traditional and civil authority. All the principal ministries have offices in each district, usually staffed with a professional officer. Local government is the responsibility of the District Council which consists of elected and appointed civil servants. The District Council maintains the smaller schools, roads, and community water supplies which are not controlled by central government.

Local development activities are coordinated through a District Development Committee, which is chaired by the District Commissioner and includes district officers of the development ministries, elected Councillors, local Members of Parliament, and party officials. All requests for development assistance must pass through these Committees, which are administered directly by the Development Division.

The Malawi Congress Party, which is the only permitted political party, plays a very significant role in rural development. The party is particularly active in rural water supply projects as there are clearly significant political benefits in supporting such a popular programme. The party organization in the villages is strong; the lower level officials are reasonably democratically elected and are generally acknowledged local leaders enjoying the respect of the community. The party is an important part of the communication network which is so essential for any large-scale rural development programme. Party officials help to coordinate and motivate communities engaged in the programme; they encourage people to attend meetings, and they participate actively in all public and committee meetings. In cases where the progress of the whole project is being hampered by the lack of cooperation of a few villages, the party is able to bring pressure to bear on the village leaders concerned to fulfil their commitment.

In addition to the party organization, there is a very strong structure of traditional authority in the rural areas. Each district is divided into a number of areas, each under the jurisdiction of a Chief. The Chief, as well as being an important figurehead, plays a key role in the administration of traditional law and custom, resolving land disputes and inter-community quarrels. Villagers are highly conscious of their allegiance to their Chief, and it is important that rural development projects take careful account of traditional boundaries.

At village level, the leader is the village headman. His principal role is to deal with minor disputes and antisocial activities, acting at the first stage of the traditional adjudication process. More serious cases are passed to the group village headman and ultimately to the Chief. The role of the headman in village leadership and decision-making varies considerably according to his personal qualities; if he has a weak personality his role may be taken over by another leader or group of leaders. Although elections for village committees are rare, leaders cannot be effective unless they enjoy a general consensus of support from the community.

These political and traditional structures form two channels of communication and authority with the village people. In general, both forms of authority are supported by the people, though often to differing extents depending on the personalities of the local leaders. There is naturally a certain rivalry between the two systems, but the competition benefits rather than hinders rural development work. The strength of these political and traditional systems is undoubtedly of great significance when considering the development of the rural water supply programme in Malawi.

As in most traditional communities, women bear the greatest burden of village life in terms of rearing children, domestic chores, and agriculture. As drawers of water they perceive particular benefits in water supply projects, and provide the major part of the self-help labour. Although the men ultimately make the decisions, it is women who form the most powerful and united pressure group, and the success of any rural development programme ultimately depends on their involvement and support.

Village size varies greatly; a small village may have 50–100 persons, a large one 1,500–2,000. In the Central Region, settlement is commonly nucleated, a cluster of houses surrounded by agricultural land, while in the south houses are generally dispersed broadly within the village boundary.

Traditionally, there is no individual ownership of land. The land belongs to the community as a whole, but is administered by the Chief through his headmen, who allocate it to private individuals. Once allocated it becomes, to all intents and purposes, private property in that the individual has rights over it to the exclusion of others.

As in many developing countries, there is a strong tradition of both mutual aid and self-help in rural communities in Malawi. For example, individual villagers frequently group together to build each others' houses; virtually all primary schools in rural areas have been built by the villagers themselves. Both the Government and the Party have maintained a major emphasis on support for self-help activities and have initiated programmes seeking a self-help response.

Water supply sector

The availability of groundwater varies considerably throughout the country. In general, average borehole yields are not sufficient for city or industrial supplies but are adequate for rural areas and smaller urban centres. Surface water, on the other hand, is abundant, with several major rivers and numerous perennial mountain streams. The situation in the dry season is eased by the natural storage provided by lakes and *dambos* (shallow basins with impervious strata at a depth of about 5 metres).

By 1977, approximately 1.4 million people, representing 30 per cent of the rural population, were served by a water supply. The three main types of supply in rural areas are deep boreholes, gravity-fed piped systems, and protected shallow wells.

The Ministry of Agriculture and Natural Resources conducts a programme for the construction of *deep boreholes* fitted with handpumps. Over 1 million people have been served by these boreholes, which serve an average of 300 people each. Approximatley 300 borehole supplies are constructed each year, at an average cost of US$4,000 (1977). Maintenance is carried out by the Ministry, with District Councils contributing towards the cost.

The Ministry of Community Development and Social Welfare developed a programme for the construction of *gravity-fed piped water systems* with the full participation of the recipient communities. By the end of 1978, 22 projects had been completed serving over 300,000 people through over 2,000 public standpipes. The mean per capita cost of projects under construction in 1979 was US$7.5; funds have been obtained through international and bilateral donor agencies. There are no house connections and the water is free. Maintenance is the responsibility of the Ministry but is carried out with the active involvement of the community.

The Ministry of Community Development and Social Welfare has also developed a programme for the construction of *shallow wells* fitted with locally manufactured handpumps, which is concentrated in the Central Region where the water table is high throughout the year. Each well serves 100 people at an estimated per capita cost of US$1.2 (1977). The Ministry provides cement, a concrete slab, the pump, and technical supervision, while the local village committees provide bricks, sand, labour, and transport. Maintenance is the responsibility of the village committees, who select a villager for training in repair and maintenance of the handpump.

In common with many developing countries, haphazard historical development led to a highly fragmented water supply sector. Until the 1970s, the activities of the various government departments involved in water supply were spread relatively thinly so that there was little overlap of responsibility. However, the increasing level of activity and new emphasis on this sector led to the realization of the need for a more rational approach. In 1980 the Department of Lands, Valuation, and Water was created and was made responsible for all rural water supply programmes.

How typical is Malawi?

While no developing country can be called 'typical', it is essential for an understanding of the programme to compare the situation in Malawi with that of other countries. In many respects, Malawi is worse off than most; it is extremely poor, with no natural resources other than its fertile land, and very little manufacturing industry; literacy is very low and there is an unusually severe shortage of trained manpower; it is a country heavily dependent on foreign aid and in an area which has experienced the turbulence and strife of the anti-colonial wars of Mozambique and Zimbabwe.

In common with many other developing countries, most of the population is engaged in subsistence agriculture. The population consists of several tribal groups with different languages. National unity is a major preoccupation of the political leadership and is one reason for the very strong political control exercised by the President.

In other respects, some of the conditions that exist in Malawi are more favourable to the development process than is the case in other countries. It is a small country with relatively easy internal communications. One advantage of the lack of mineral resources or industry is that the country has avoided the intense social problems of urbanization that are so common in other developing countries, with the result that the traditional rural society is still strong. Because of the serious shortage of trained professional manpower at independence, many of the former colonial expatriate civil servants were asked to stay on, and have only been replaced gradually as professional Malawians become qualified for the posts. The Government has continued to recruit expatriates in the Civil Service where no qualified Malawian is available. Thus Malawi has avoided the enormous upheaval that inevitably takes place when all expatriates have suddenly been replaced after independence by untrained and unqualified nationals. As mentioned above, Malawi is less plagued by excessive bureaucracy than many other countries. This is not to say that bureaucratic problems do not exist—the rural water supply programme had to overcome many formidable problems of this nature. The difference may be that, in Malawi, the bureaucracy does eventually respond to continued perseverance. (Perseverance in these matters is an important reason for the successful way in which the programme in Malawi slowly developed.) In having a single, strong political party, Malawi is fairly typical of a significant number of developing countries. Where it may differ from others is that the strong political discipline, which can be put either to good or bad use, is used to very good effect in the field of national development. In other countries, many right-wing governments are not as development-oriented as in Malawi, whereas many left-wing governments have not been able to exert the same degree of political discipline and authority to ensure the success of their development policies.

To conclude this chapter, it is important to stress that, whereas an appreciation of the situation in Malawi is important for an understanding of how the programme developed, the real value of the experience which the author endeavours to bring out in this book is that there are certain principles and guidelines which are not specific to Malawi and which can be applied in most developing countries. As evidence of this, Nepal, where the author is currently working, is utterly different from Malawi in terms of geography, history, politics, social organization, and culture. It is obvious that the programme in Nepal should not be conducted in exactly the same way as in Malawi. Yet the *application* of the fundamental principles and guidelines derived from the Malawi experience to the *specific* situation that exists in Nepal shows where the programme in Nepal went wrong, and indicates the direction in which the programme should go. Such an analysis is an essential step towards improving the performance of existing programmes in all developing countries in pursuit of the goals of national development and of the International Drinking Water Supply and Sanitation Decade.

CHAPTER 2
Historical development

2.1 Introduction

The account that follows is given for two reasons. First, it is a factual record that should be of interest to planners and managers of similar programmes, because there are so few such records available. Secondly, the Malawi programme has attracted attention as one of the more successful programmes, particularly in the fields of community participation and low-cost technology. The experience of Malawi is, therefore, of significant importance for the planning and implementation of such programmes in other countries. To understand the factors leading to this, it is essential first to examine its historical development from its inception in 1968.

To facilitate the understanding of this account, the principal events are listed in chronological order in Table 2.1. Figures 2.1 to 2.3 record the growth of some key parameters with time.

2.2 Pilot Phase

It is significant that many development programmes may originate in unplanned events whose consequences are not foreseen at the time. The programme in Malawi was not planned in advance; there were initially no master plans, feasibility studies, or even project documents. On the contrary, the programme was born at village level as a simple response to the expressed needs of a small community. From that point it evolved as a logical sequence of events but remained a village-based rather than a planner-based programme. In retrospect, and for the sake of analysis, this early period, in which two projects were completed, can be described as the Pilot Phase.

Chingale—the first project

In 1968 the Ministry of Community Development and Social Welfare (MCDSW) undertook a community development project covering 16 villages in an area called Chingale in the Southern Region of Malawi (see Figure 2.7). The objective of the project was to concentrate the ministry's inputs into a small area and conduct adult literacy classes, stimulate income generation for women, support the self-help construction of primary schools, health posts, etc.—these being the "traditional" activities of the ministry.

To assist the people to identify such construction projects, the ministry sent an engineer who had recently joined the staff. From his discussions with the project

Table 2.1 Chronological development of the rural water programme in Malawi (1968–79)

1968	Ministry of Community Development and Social Welfare (MCDSW) selects Chingale as a Community Development Project Area. Water identified as main problem
1968–69	Construction of Chingale Pilot Water Project (pop. 3,000)
1969–70	Construction of Chambe Project (pop. 30,000) in Mulanje District
1971	Formation of Water Projects Section (WPS) within MCDSW
1971	Appointment of Assistant Engineer and transfer of 3 Community Development Assistants to WPS
1971	Preliminary work on Mulanje West and Phalombe Projects
1971–74	Construction of 5 small projects in all three Regions
1972	Selection and training of 20 Project Assistants—appointed to Mulanje West (15) and Phalombe (5) for in-service training
1972	Commencement of Mulanje West (pop. 75,000) funded by UNICEF
1973	Commencement of Phalombe (pop. 90,000) funded by DANIDA
1974	Selection and training of 10 Project Assistants—appointed to Phalombe for in-service training
1974	Expansion of programme to new areas. Experienced staff transferred to conduct demonstration projects in focal areas
1975	Mulanje West complete—more staff transferred to Phalombe and other projects
1975–77	Introduction of operation and maintenance procedures for Mulanje West
1975–76	Two Technical Officers appointed to projects in Southern Region (1) and Northern Region (1)
1976	Selection and training of 20 Project Assistants
1976	Commencement of Namitambo (pop. 60,000) funded by DANIDA and Sombani (pop. 40,000) funded by ICCO (Netherlands)
1977	Completion of Phalombe—introduction of operation and maintenance procedures
1977	Commencement of Zomba East (pop. 100,000) funded by CEBEMO (Netherlands)
1977	Appointment of 2 engineers—1 to take over as Project Manager, Mulanje, 1 as Project Manger, Zomba
1978	Two Technical Officers, after 2 years field training, sent overseas for university degree courses
1978	WHO/World Bank Sector Study Report
1979	Commencement of Mulanje South (pop. 45,000) funded by CIDA. Recruitment and training of 24 Project Assistants
1979	New career structure for field staff, changing over to technical grades
1979	Decision to transfer WPS from the MCDSW to the newly-created Department of Lands, Valuation, and Water

Historical development

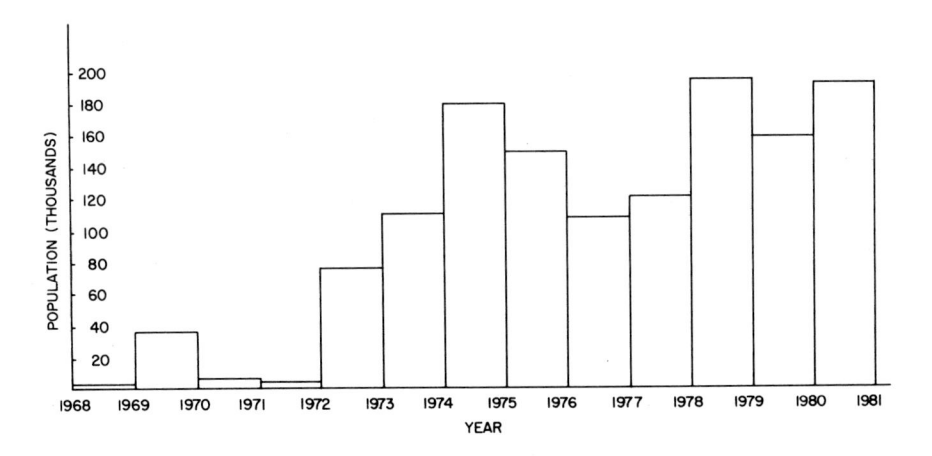

Figure 2.1 Population actively engaged in project construction in Malawi in each year

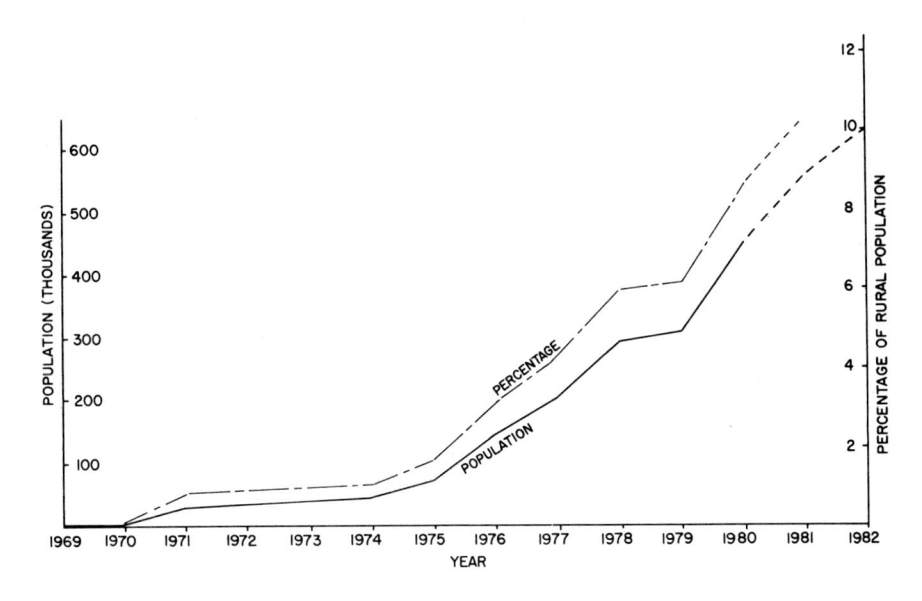

Figure 2.2 Cumulative population served by programme in Malawi

committee it became clear that the major problem facing the community was their water supply. In the wet season they used shallow, hand-dug, unprotected wells, while in the dry season the women had to walk several miles to dig in the river bed.

At the engineer's request, the committee took him up to a perennial stream flowing off the lower slopes of the mountain that bordered the area. It was clearly feasible to build a small intake and pipe water by gravity to the villages. The committee was sceptical but agreed to mark a route for the pipeline as they walked

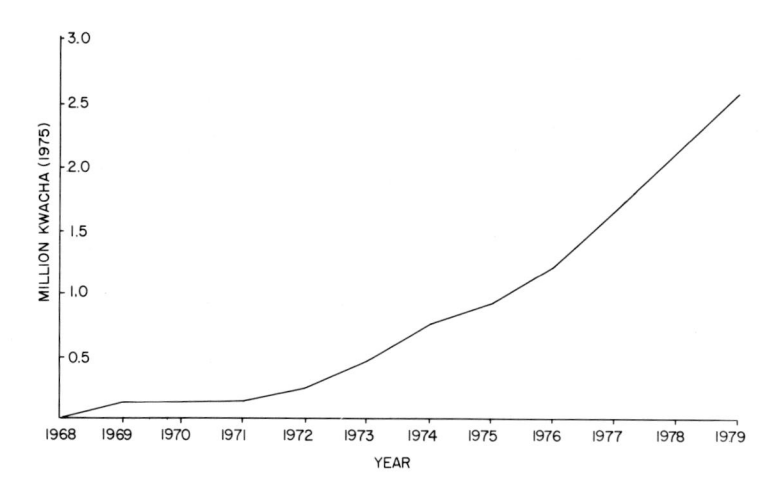

Figure 2.3 Cumulative programme expenditure in Malawi, adjusted to 1975 prices

back to the village. At a committee meeting soon after, the engineer stated that the Government would provide pipes and materials if the people would dig the trench. The committee agreed to this proposal.

The engineer drew up the design and schedule of materials, the ministry secured the funds from a donor and purchased the necessary pipes and materials. With the help of the Community Development Assistants (CDAs) who were also working on the project, the committee organized the villagers into a programme for digging the trench. Once the trench was dug, they laid the pipes (PVC), backfilled the trench, and finally installed their public standpipes.

This pilot project was completed within 1 year at a cost of US$7,500 (1969); 25 km of PVC pipe were installed, supplying water through 25 public standpipes to 3,000 people at 25 litres per head per day. (The supply was later extended to serve a further 2,000 people.) It is still functioning today, being operated and maintained by the villagers with technical and material assistance for the Government.

Chambe—the second project

There was another focal area of community development activity in the neighbouring district of Mulanje (see Figure 2.7). The popular success of the first project led the ministry to consider the possibility of conducting a similar, but larger project in this district where the inhabitants were known to be suffering from an acute lack of water.

Mulanje, with a total population of 500,000 in about 700 villages, is dominated by a 3,000 m high mountain, which is the source of several perennial rivers and streams. Despite the abundance of water flowing off the mountain, the water supply situation for the majority of the population living on the plain below was very poor. In the wet season, surface water and shallow wells became badly polluted as the result of a relatively high-density population with virtually no sanitation, and in the dry season the water table dropped so that streams and wells dried up completely.

Historical development

Figure 2.4 The traditional water sources are gradually drying up

(Photo: the author)

The water supply situation had become progressively worse in recent decades with the increase in population raising overall demand, and the increase in cultivation leading to greater run-off and a lower water table. As wells dried up, communities were forced to migrate towards the larger rivers of the plain, where water could still be found in the dry season by digging in the river bed. The Government had responded to the problem by installing some boreholes, but these were very few and the water was saline.

The Chambe area on the west side of the mountain had been receiving assistance from the ministry in the form of an adult literacy programme and support for the construction of some primary schools by self-help. One of the larger rivers in the district flows down the mountain slopes adjacent to this area, but abruptly turns south and bypasses most of the population living on the plain. The ministry engineer, who knew the area well, conducted a brief survey and showed that a gravity water system could be constructed from this source to serve about 30,000 people in 60 villages.

The engineer and five CDAs were posted to the area to supervise the project. In order to demonstrate the feasibility of the project, leaders from the area were first taken to the completed project at Chingale. They were shown around by the water supply committee who explained how the work had been carried out. The leaders then returned to their own villages and were given time to discuss their visit within their own communities. Once public interest was aroused, the Chief called a public meeting at which the Government offered to help if the people would provide the labour.

Committees were formed, and the work got off to an enthusiastic start. The villagers dug the main line from the intake, and then started work on their branch lines. The intake and sedimentation tank were constructed; the PVC pipes arrived and were distributed to dumps along the lines. However, this stage coincided with a period of political unrest in some parts of the country and the project itself became the victim of political discord. The self-help labour programme almost stopped. Fortunately, however, project staff were able to maintain a low level of self-help activity, particularly at the point where water was already flowing out of the pipeline. The political unrest died down and the project had enough inherent momentum and public support to recover from the set-back. It was completed in 1970 and involved the installation of 95 km of PVC pipe delivering water through 180 public standpipes to 30,000 people at a total cost of US$80,000 ($2.67 per head).

Lessons of the Pilot Phase

There are a number of lessons to be drawn from this Pilot Phase which have important implications for the planning and implementation of similar programmes in other countries. On the positive side, the Government had proved for itself and for potential donors that reasonably large-scale piped water supplies could be installed in rural areas by self-help labour. It proved to the people that the Government was sincere, enabling them to overcome the traditional scepticism that all rural people have for government programmes. It also demonstrated to the people a fact that most of them would never have believed possible—that water could flow very long distances inside a pipe.

However, a number of lessons were learnt as a result of the problems experienced on the second project. These concerned the size of the project, field staff, supervision, and the role of self-help. With the benefit of hindsight it is clear that the second pilot project was the "critical point" in the development of the programme, and it is worth looking in more detail at the lessons learnt.

Project size

As the initial enthusiasm on the project waned it became apparent that there was a credibility gap between what the leaders saw when they visited the first project and what they were able to convey to the people. These natural doubts, based on the traditional conservatism of rural people, were exploited by clandestine political elements opposing the Government. Rumours were spread about the Government's motives in supplying water, and there were some isolated violent incidents. The people either became genuinely disaffected, or were too scared to continue to participate in the project.

Such political rivalries and tensions are experienced by all programmes to a greater or lesser extent. The lesson here is that the project was *too large* for a pilot project. A smaller project would have been less vulnerable to exploitation because there would have been closer communication between project staff, community leaders, and the people. Once a small pilot project has been successful, larger projects are likely to receive greater popular support and have a greater chance of surviving such political tensions.

Appropriate field staff

The experience also shows up the importance of having the right type of field staff for the project. The performance of the CDAs indicated that, with a few exceptions, they were not generally suited to the technical and managerial tasks involved. After 2 years of academic training in a modern training college, they had difficulty in identifying themselves with the problems and aspirations of the rural communities from which they had come. They felt themselves attracted to 'white-collar' administrative work and were reluctant to engage in more physical tasks, such as helping the people move a rock or join a pipe. This phenomenon is fairly common to rural development programmes in most countries.

The lessons are that field staff (a) should be drawn *directly from rural communities* and (b) should receive a *short period of appropriate training* that would concentrate on technical operations and the organization and management of village labour.

Supervision

The experience of both pilot projects showed the necessity for closer supervision to ensure an adequate technical standard. Pipe trenches were rarely straight and rarely dug to the correct depth. Standpipe aprons and drains were of a poor quality and sometimes non-existent. It was also clear that greater attention would have to be paid to reduce erosion along the pipeline. *A greater number of field staff* would be needed on each project to ensure a higher technical standard and greater attention to detail.

Figure 2.5 **A tapstand on one of the pilot projects—the design was later improved to include apron, drain, and soakaway**

(Photo: unknown)

Figure 2.6 The foundation of a tapstand apron and drain ready for concrete, as developed later in the programme

(Photo: H. Van Schaik)

Role of self-help labour

The experience of the Pilot Phase showed that self-help is suitable for unskilled work, such as trench-digging and backfilling, excavation of tank sites, collection of materials, carrying pipes, and even joining them. However, the construction of the intake, tanks, and other concrete works was not well done. The project committee had selected some local artisans and had collected money to pay them. It was difficult to ensure an adequate quality of work as the work required was beyond the experience of most of the artisans. In addition, there were problems with the collection of funds and accounting for them. It therefore became clear that such construction work would have to be done by a team of trained builders who would gradually accumulate experience, and who would have to be paid from project funds.

Maintaining project momentum

A significant reason why the second pilot project overcame the problems is that work was never allowed to stop completely. Work on a self-help project generally starts with great enthusiasm and it is important to ensure the initial momentum is maintained as long as possible. This means that the labour must be well-utilized, so that the people feel they are making progress. Supplies of pipes and fittings and construction materials must be on site in time so that there are no delays in which community enthusiasm will fall. There is also a tremendous psychological impact if the villagers can see the water flowing even while work is proceeding, so the intake works must be completed early on. Field staff also need to be aware of the importance of keeping the work going, and need to make special efforts if the enthusiasm of the community falls.

Action resulting from the Pilot Phase

One of the main objectives of a Pilot Phase is to test the institutional arrangements for the programme. As a result of the first two projects in Malawi, the ministry decided to develop a more specific structure to replace the *ad hoc* arrangements that had been made so far. From this point, the activities in rural water supply began to develop into an established, rational programme.

A Water Project Section (WPS) was created under the control of a Senior Water Engineer, who was appointed to be Project Manger in Mulanje District, where the initial effort of the programme was to be concentrated. In addition, a second engineer was recruited as assistant to the Project Manager and three CDAs who had proved themselves in early projects, were appointed as Supervisors.

Most significantly, the Project Manager was also authorized to recruit and train 20 field staff, known as Project Assistants, and to train a suitable local builder to form a construction team specializing in intakes and tanks.

2.3 Consolidation Phase

The poor performance of some rural water supply programmes in developing contries is often largely due to the fact that they expanded too quickly after a brief Pilot Phase. In their haste to develop as rapidly as possible in response to the political demand for water, these programmes have not been able to consolidate their experience, develop manpower, and establish standard routines and procedures. A period of consolidation is absolutely essential for the healthy development of a programme; a significant reason for the success of the programme in Malawi is that the next two projects were used for this purpose and can be described as the Consolidation Phase.

The first major projects

Following the successful completion of the pilot projects, the ministry drew up plans for two large projects in the same district. Mulanje West Project was designed to serve two Chiefs' areas with a combined population of 75,000 in 120 villages. Phalombe Project was to serve another Chief's area with a population of 90,000 in 135 villages (see Figure 2.7).

Although these projects were large for a relatively new organization, the ministry had three reasons for feeling confident. First, it had gained confidence in itself from the Pilot Phase, and felt that the lessons learnt would enable problems in future projects to be avoided or resolved. Secondly, there was now provision for an adequate number of appropriately trained field staff. Thirdly, and most important, it felt that this time it was genuinely responding to the demands of the people, who no longer needed to be persuaded or convinced. The demonstration effect of the Pilot Phase had convinced the people that they also wanted standpipes in their own villages, and that the Government would help them. Thus the people were now

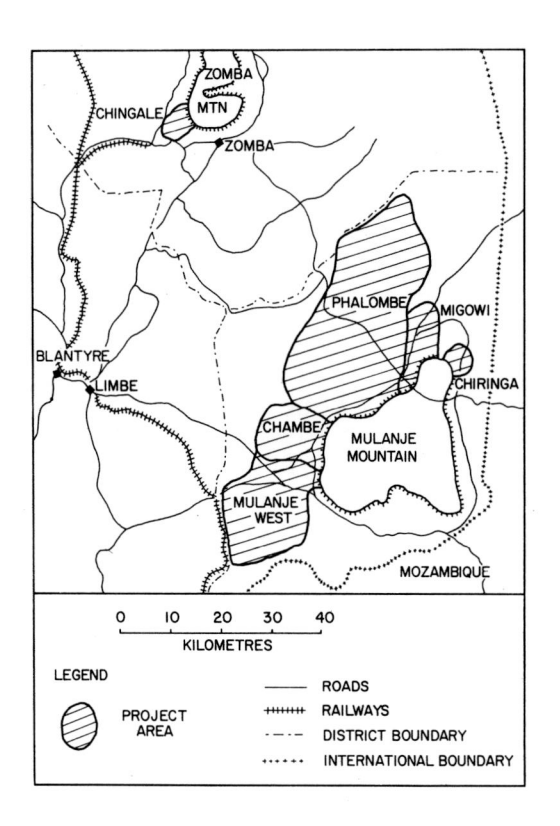

Figure 2.7 Map locating Chingale and early projects in Mulanje District, Malawi

urging their leaders to put pressure on the Government, which was a healthy situation.

Preliminary work was carried out under the three Supervisors, while 20 Project Assistants were recruited and trained by the engineers. It was decided to concentrate first on Mulanje West until the project was well under way, and then to transfer some staff to Phalombe. The organization and technical details of these two projects are described in general terms in Chapters 4 and 6. From a historical point of view their significance lies in three major developments that strongly influenced the character of the subsequent programme.

The first was the development of a *strong team consciousness* among project staff, with close communication between Project Manager, Supervisors, and Project Assistants. The team spirit was generated partly by the field training course, and partly by the experience of working together. This was an important factor in the motivation of field staff.

The second was the development of *standardized procedures and techniques* for all the operations carried out by field staff. The main bulk of experience was gained during this time and procedures were tested and improved by the team as a whole, so that the results represented the collective experience of all staff.

The third development, closely related to the above, was the *improvement of the technical standard of installation*. Trenches were now dug straight and to the correct

Historical development

depth, standard designs were laid down for standpipe aprons and drains, measures were adopted to reduce the erosion of pipelines, etc.

These three developments are key elements in the Malawi experience, and will come under closer study in later chapters.

Details of other major projects that followed are given in Appendix 1. All major projects so far constructed have been in Mulanje and two neighbouring districts. Together they account for 75 per cent of programme expenditure and 70 per cent of the total population served by the programme.

2.4 Expansion of the Programme

One of the effects of a successful Pilot Phase in any programme is likely to be a growing demand to spread to other areas of the country. In Malawi, requests came in from District Development Committees in the Northern and Central Regions. The ministry decided to carry out two small projects in the Northern Region, one in the Central, and two in new areas of the Southern Region. This was a reaction to the demand, rather than a declared policy of expansion. The main effort of the programme continued to be concentrated in the two major projects in the South.

Policy for expansion

As these smaller projects were completed, the demonstration effect stimulated still further demands from District Development Committees all over the country. At this point, many programmes have yielded to political pressure and have tried to expand rapidly, without an adequate manpower base or reservoir of experience, and without having developed the procedures and techniques that result from a Consolidation Phase. In Malawi, this political pressure was resisted by the ministry who decided to follow a policy of controlled expansion. The main constraint was manpower, which was limited in both numbers and experience. Even if new field staff could be recruited, the engineers advised that they would need at least 2 years experience on a major project under close supervision before they would be able to supervise a small project on their own. Another concern was that the "demand" from all over the country was not the genuine popular demand of villagers who had actually seen a project, but was the institutionalized demand from District Development Committees. For a programme relying on self-help labour, it is essential that the demand genuinely exists at village level, which can only come about as a result of demonstration projects.

It was therefore decided to concentrate the expansion of the programme on a few focal areas of outstanding potential, rather than spread the programme thinly all over the country. Small demonstration projects were conducted in these focal areas. In 1974 the first Project Assistants, who had by then gained 2 years experience on the major projects, were appointed to six small demonstration projects in the three Regions. These smaller projects were funded from a 'rolling fund' with donor support, enabling new projects to be undertaken in adjacent areas as the demonstration projects were completed. In 1975 the ministry appointed a Technical Officer (Diploma Engineer) to one focal area in each of the three Regions. (Details of minor projects are given in Appendix 1.)

2.5 The Role of Donor Agencies

As donor agencies play an important part in the rural water programmes of many developing countries, it is significant to record the role that they played in Malawi.

The first project was supported by the United States Agency for International Development (USAID), from a special fund that supported small self-help projects. This is an example of the seminal role that a donor can play in helping to initiate an activity that later develops into a full-scale programme.

The second project was funded by a British charity, OXFAM. The financial commitment was relatively high for this organization and there was undoubtedly an element of risk in supporting a new type of project involving so many people in a new area. In fact, as discussed above, the second project turned out to be the critical point in the development of the programme and the support of this donor proved to be crucial. It was as a result of this project that the programme began to take shape; this is an example of how a small donor can act as a vital catalyst in the development of a programme to the stage when it can attract major donors, or be taken over by government.

The United Nations Children's Fund (UNICEF) and the Danish International Development Agency (DANIDA) were the funding agencies for the two major 'consolidation' projects, Mulanje West and Phalombe. The former was attracted by the potential health benefit to children, the latter by the agricultural development that would result from the provision of domestic water to encourage settlement in a relatively undercultivated area.

For the expansion of the programme UNICEF and the Christian Service Committee of the Churches in Malawi (CSC), a Malawi-based organization, combined to provide a 'rolling fund' for as many small projects as could be undertaken each year. UNICEF provided the pipes and all materials that had to be imported, while CSC provided the local costs.

In response to the success of the programme, and because of increased interest in the development of rural water supplies, several other major funding agencies have since become involved. The agencies involved are listed against their respective projects in Appendix 1.

2.6 Conclusion

Several aspects of this historical account will be brought out in later chapters. There are two main points that need to be stressed here, as they have particular relevance to programmes in other countries.

The nature of demand

It is a fallacy to assume that, because a community is obviously in need of an improved water supply, it will automatically provide voluntary self-help labour. Rural communities have many priorities, of which water may be only one. Many programmes experience problems because the self-help commitment fails, despite the fact that the traditional and political leaders, and bodies such as local development committees, have expressed to the government the urgent need for a water supply in their area. Such demand may be called institutionalized demand—the reaction of the more educated, or politically opportunist, leaders to the possibility of securing government help for the communities they represent. Very often the people themselves are not aware at all that their representatives have asked for a water supply.

In the Pilot Phase, or in a demonstration project in a new area, this is the type of 'demand' to which the programme has to respond. If the people have not seen a village water supply, they will not really believe that it is possible, whatever the leaders may say. In these circumstances the implementing agency has to guide the pilot or demonstration project through to a successful conclusion, being particularly sensitive to the people's scepticism or lack of real enthusiasm. Political or community rivalries are likely to be a major problem in these circumstances, as happened in the second pilot project in Malawi.

Some programmes never get out of this situation—responding only to this institutionalized demand and repeatedly experiencing problems with community participation. The only way a self-help programme can really be successful is to stimulate a genuine popular demand, so that the people themselves are committed to the project, are prepared to contribute their labour and overcome the many problems that they will encounter. This genuine demand is most effectively stimulated when the people actually witness a project being well-constructed nearby—a process that may be called the demonstration effect.

The programme in Malawi succeeded in stimulating this genuine popular demand, by carefully guiding the pilot projects through to a successful conclusion and then responding to the demand of the neighbouring communities.

Controlled expansion

In most developing countries the biggest single constraint to the development of a programme is the shortage of trained manpower and the consequent inability to provide adequate supervision. Despite this, many governments have been unable to resist the political pressure to expand a popular programme such as rural water supply, often with disastrous results. In these cases poorly trained manpower with virtually no experience of the programme are forced to implement water supply projects with no supervision, no procedures, and no support. It is hardly surprising that such projects run into difficulties, and often fail; the demonstration effect on the people is counterproductive and yet the politicians clamour for more projects, and so the cycle goes on.

In Malawi the ministry did manage to resist this political pressure to a certain extent, and did not expand the programme until field staff had been sufficiently trained and had developed enough experience on the two major 'consolidation' projects. Experienced staff were then posted to a few focal areas specifically chosen for their high potential for similar rural water supply schemes, to implement demonstration projects, and later to implement larger projects in neighbouring areas. Far from delaying the development of the programme, this policy has enabled Malawi to develop a strong programme and to become one of the few developing countries that is likely to approach the Decade target of water for all by 1990.

CHAPTER 3
Programme organization and management

3.1 Introduction

The problems most likely to be experienced in a rural water supply programme are not usually of a technical nature, but are generally due to poor organization and management. The management functions can be divided into those that are carried out at central or headquarters level and those that are performed at field level. This chapter discusses the central level.

3.2 Objectives

Rural water supply programmes generally develop in response to public demand. From the villager's point of view the principal objective is to reduce the burden of water collection by providing a water supply within a reasonable distance of the home. From the government's point of view the principal objective is to satisfy as many people as possible. Whatever other objectives may be desirable, the pursuit of these objectives will inevitably remain the fundamental motivating force behind the programme. This is as it should be, for a successful programme must be village-based rather than planner-based.

However, it is also true that there are potential benefits other than those perceived by the villagers, the principal one being the potential health benefit. In most developing countries, diseases transmitted as a consequence of inadequate water supplies and sanitation are highly prevalent and significantly hinder economic growth. Therefore, it is clearly in the national interest to maximize the potential benefits that may occur from an improved water supply. This should be an important secondary objective of the rural water supply programme. For many donor agencies this may be the principal objective, but donors should understand that this is unlikely to be the case either for the government or for the people.

3.3 Organization of the Programme

Specialist approach

Rural water supply programmes need to develop a unique approach, blending technical competence with social skills, which can only be achieved within a *specialist unit*. Even within such a unit it takes several years to develop a team of experienced

1970

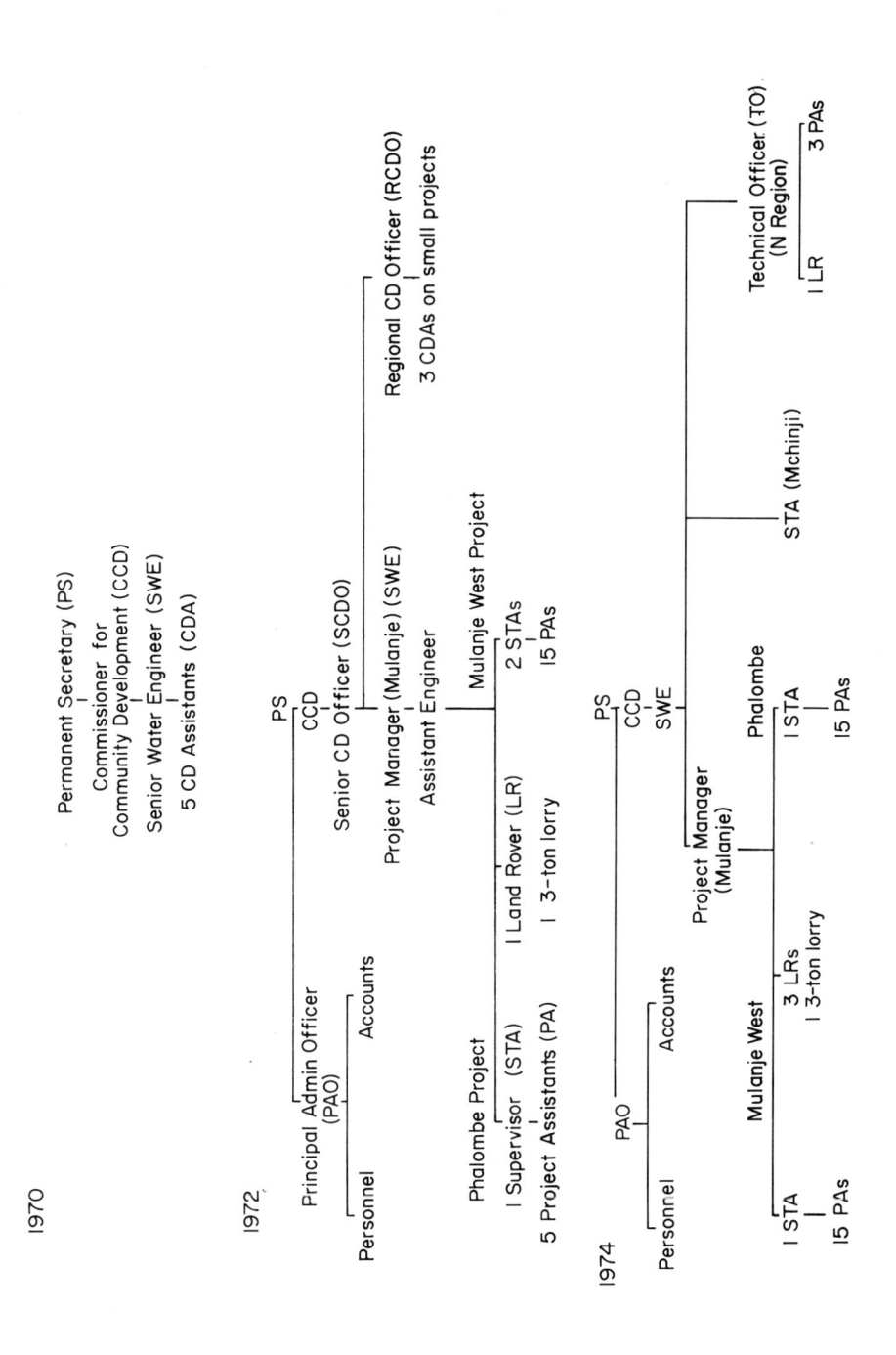

Permanent Secretary (PS)
|
Commissioner for
Community Development (CCD)
|
Senior Water Engineer (SWE)
|
5 CD Assistants (CDA)

1972

```
                                PS
                                |
                               CCD
                                |
                    Senior CD Officer (SCDO)
                                |
    ┌───────────────────────────┼──────────────────────────┐
Principal Admin Officer    Project Manager (Mulanje) (SWE)    Regional CD Officer (RCDO)
      (PAO)                           |                              |
  ┌────┴────┐              Assistant Engineer                 3 CDAs on small projects
Personnel  Accounts                   |
                          ┌───────────┴───────────┐
                    Phalombe Project         Mulanje West Project
                          |                        |
                  1 Supervisor (STA)           2 STAs
                  5 Project Assistants (PA)    15 PAs
                  1 Land Rover (LR)
                  1 3-ton lorry
```

1974

```
                                PS
                                |
                               CCD
                                |
                               SWE
                                |
                ┌───────────────┼──────────────────────────┐
              PAO        Project Manager            Technical Officer (TO)
          ┌────┴────┐       (Mulanje)                    (N Region)
      Personnel  Accounts      |                              |
                    ┌──────────┼──────────┐              ┌────┴────┐
              Mulanje West  Phalombe   STA (Mchinji)   1 LR    3 PAs
                    |          |
                 1 STA       1 STA
                 15 PAs      15 PAs
                 3 LRs
                 1 3-ton lorry
```

Programme organization and management

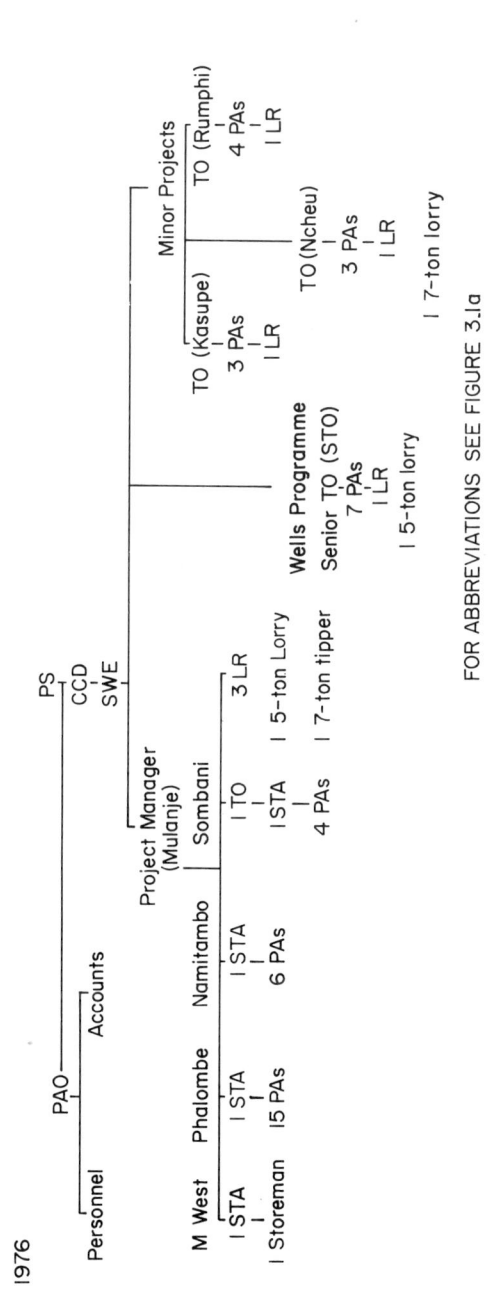

Figure 3.1 Organization charts of Water Projects Section in Malawi as at 1970, 1972, 1974, and 1976

manpower with the appropriate skills—in a non-specialist unit it is impossible. Some countries utilize so-called "multi-purpose" technical workers who are expected to be able to build water systems, bridges, roads, and buildings. This approach is disastrous, as the chronic shortage of trained manpower in most developing countries rules out any possibility of adequate supervision.

The programme should therefore have its own organization, with permanently appointed staff who are not subject to transfer to other programmes every few years.

Parent ministry

The unit may be either in a technical or non-technical ministry. The advantages of the former are that the unit will be able to draw on the technical resources of the parent ministry, for example for procurement of supplies or the sharing of workshop facilities. However, the traditional engineering ministries, such as the Ministry of Works or its equivalent, generally have a highly technocratic and urban-oriented approach, and have little experience of working with simple technology and with rural communities. For this reason it may be preferable to be located within a rural-oriented ministry, such as Rural Development, Community Development, or Agriculture, even if this means less technical support and less understanding on the part of the parent ministry of certain technical aspects of the programme. The situation will vary from country to country, but in general, the degree of rural and community orientation is more important than the technical orientation.

Technology-specific sub-programmes

In many countries haphazard historical development has resulted in a multiplicity of government departments being involved in the construction of rural water supplies. Each department may be responsible for different types of water supply and may operate in different geographical areas—for example, the Department of Geological Survey may be responsible for groundwater investigation and installation of deep boreholes, the Ministry of Works for dense rural communities and district centres, the Ministry of Health for shallow wells, the Ministry of Rural Development for piped supplies for rural villages. Each of these will have their own manpower, procedures, and expertise which have been developed over the years. Although it is desirable to bring all these specialist units into one parent ministry, they should not lose their identity, but should remain as distinct sub-programmes of the whole.

Programme organization in Malawi

In Malawi the Water Project Section was set up as a self-contained unit within the Ministry of Community Development and Social Welfare. The growth of this section from 1970 to 1978 is shown in Figures 3.1 and 3.2. The Project Manager of each major project and the Technical Officers of each focal area (who supervise a number of smaller projects) report directly to the Senior Water Engineer at ministry headquarters. This organization was highly centralized and led to a high order of operational and managerial efficiency within the section. The Senior Water Engineer

Programme organization and management

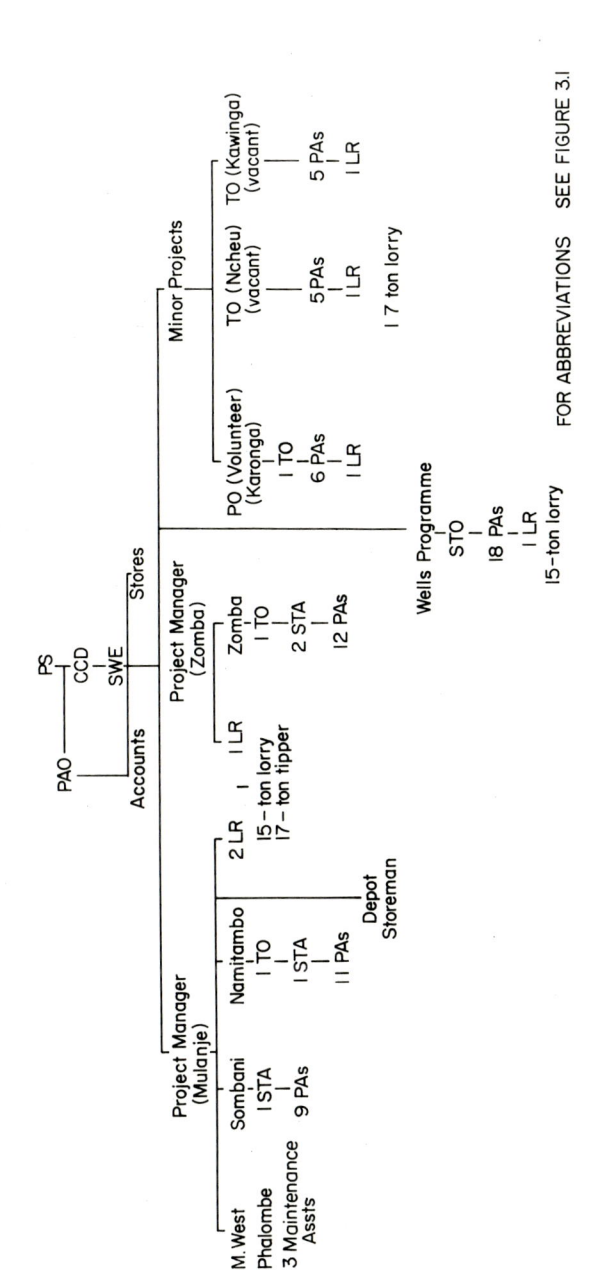

Figure 3.2 Organization of Water Projects Section in Malawi as at 1978

maintained a very close link with field staff at all levels, who were themselves very conscious of his support at the ministry when problems arose. The chain of authority worked principally by personal communication and there was a minimum of paperwork. This was possible within a relatively small organization and was a major factor in the successful development of the programme. As the programme grows, however, it becomes necessary to break down into smaller units, with intermediate levels of authority at, say, Regional level. In large countries, or countries where communications are particularly difficult, more decentralization may be necessary, but this should only come about when the programme has developed enough manpower and experience to expand gradually to each new area.

The Water Project Section in Malawi initially concentrated on gravity-fed piped water schemes, for which there is enormous application in the country. Later, the section started a sub-programme to construct shallow, hand-dug, protected wells in parts of the Central Region, where this technology was the most appropriate. This has remained a separate sub-programme, with its own staff of Project Assistants, Supervisors, and engineers.

3.4 Programme Planning

Apart from day-to-day management relating to the execution of current projects, the head of the programme should be concerned with future development, working on proposals for new projects, estimating costs and manpower requirements, and producing formal project submissions.

Applications for new projects

In a self-help programme it is essential that the community itself makes the formal application for a water supply system. Requests originating from the community are often first made informally, by a member or group from the community concerned, either to the local development committee, or direct to the office of the implementing ministry. Such a request may not necessarily be made with the full knowledge or support of the community and it is wrong to treat this as a formal application. Instead, the community leaders should be asked to fill out an application form which is so designed that it requires a certain amount of serious consideration, yet not so complicated that it cannot be filled in easily. Ideally, the application form should be taken to the community either by a member of the local development committee, or by a member of the local administration, who should help the community leaders fill out the form and ensure that it generally reflects the views of the leaders and not just of a few individuals. The completed form should then be submitted to the local development committee who should place the requests in order of priority. The ministry may invite a certain number of applications from each committee each year.

Initiative for new projects

While it is essential for the community itself to make the formal application, the initiative for the request may originate from other sources. For example, the implementing ministry may see a potential project as a logical extension of previous work; the Ministry of Agriculture may identify areas of agricultural potential where the lack of domestic water is a constraint to settlement; the Ministry of Health may identify areas with a particular health problem, such as schistosomiasis or trachoma, which would benefit from water supply and sanitation; an integrated rural development project may include water supply as an essential component of rural infrastructure; potential areas may be identified by the Ministry of Water Resources as a result of special studies or master plans. Frequently the initiative originates from political or influential persons who are keen to bring benefits to their particular community.

Whatever the source of the initiative, all such requests should be referred back, through the local development committee, to the community itself who should be invited to take over the initiative by submitting an application form. In this way, it is quite clear to all that the community is actively participating in the request procedure. Without this, community participation is meaningless, and the project is likely to run into serious problems.

Feasibility surveys

Applications that do not meet the selection criteria (see next section) should be rejected and the communities should be informed accordingly. Feasibility surveys should then be conducted for the remaining valid applications. These surveys should determine not only the technical, but also the social feasibility of the project. The technical feasibility may be relatively easy to assess, but the social feasibility is a more complex problem. It is *essential* that the feasibility survey team includes personnel who are experienced in the programme and who know what to look for within the community. There is no need to incorporate special community "facilitators" or "communicators" into the team; if the survey team members have several years experience they will themselves be expert communicators and will understand the issues much more clearly than a non-technical person. Forms and questionnaires are useful aids, but can never replace experienced judgement. The feasibility survey is probably the most important stage of the project; yet many programmes have got into serious difficulties by using inexperienced staff for surveys and undertaking projects that are technically sound but socially infeasible or highly complex.

Procedure for feasibility survey

It is important that the community is given adequate warning of the arrival of the feasibility survey team. The team should also preferably be accompanied by a district level official who knows the area and some of the leaders. If the date of arrival is known, the leaders should be asked in advance to call a public meeting. If the date is not certain, a public meeting should be called on arrival, and adequate time allowed for villagers to gather. At the meeting the local leader should introduce the survey team to the community and explain why they have come. The team should then generate a discussion on the community's water problem by asking

Figure 3.3 Conducting a feasibility survey of a river intake
(Photo: L. H. Robertson)

suitable questions—for example, "Which area needs water most? What are the possible sources? Would the people be willing to provide voluntary labour?'. The meeting should then discuss the necessary arrangements to conduct the technical

Figure 3.4 A public meeting in which a feasibility survey team discusses its findings with the villagers and explains what will happen next
(Photo: UNICEF, Nepal)

part of the survey and a group of villagers should be selected by the meeting to accompany the survey team. The technical part of the survey will usually have to be conducted the following day.

After the technical part of the survey is complete, the team and the group of villagers should report back to the community at a final public meeting. They should describe how the survey was carried out and discuss the most feasible solution. The survey team should explain how the feasibility survey will be processed and should stress that the government cannot make any commitment until all feasibility surveys have been completed.

Experienced team members will be able to assess the social feasibility by observing the enthusiasm of the community and the degree of participation in the meetings and the survey; by judging how much popular support the leaders have; by discussions with leaders and ordinary villagers; by ascertaining past performance of self-help projects, if any; and by assessing the possibility of disputes over, for example, the water source or the pipeline route.

The feasibility survey, if properly conducted, should take at least 2 days, and may take many more depending on the size of the area.

3.5 Project Selection Criteria

Geographical

It was stressed in Chapter 2 that the programme should adopt a policy of limited expansion based on certain focal areas. Applications should not be invited, therefore, from communities or local development committees out of reach of these areas.

Maximizing water resources

Requests from the local development committee often concern a single village or small group of villages that have been most active in presenting their case. The feasibility survey, however, may show that the source indicated can serve a much larger area. It is almost certain that, once water is supplied to the original villages, the adjacent villages will ask for water from the same source. Unless provision is made in the original design for an adequate pipe size, the whole line from the source will have to be laid again. In some cases, depending on the nature of the communities themselves, it may be advisable to supply only the original villages in the first phase, but make provision to extend to adjacent villages. However, whenever possible, it is preferable to include the adjacent villages in the original project as this is much more efficient in terms of manpower, time, and cost.

Cost

Per capita costs will obviously vary considerably from project to project, and the cost criterion should not be applied too strictly. Many of the more expensive projects may be fully justified on other grounds. In most developing countries, even the more expensive rural water supply schemes involving self-help labour will have much lower per capita costs than an urban scheme, mainly because of a lower level of service. Nevertheless a maximum per capita cost criterion should be applied, so that excessive funds are not utilized to benefit only a few people. This maximum will vary enormously from country to country. In Malawi there was no rigid maximum, because the costs were so low. One of the major 'consolidation' projects was three times the per capita cost of previous projects, but was still cheap at about US$7 per head (1973).

Potential benefit

The actual benefits of a water supply are difficult to quantify. Nevertheless, an assessment of potential benefits should be made during the selection process. The most obvious situation is where there is a desperate need for water in a seasonally very dry area which has a substantial population. In this case the chief benefit would be the reduction in the burden of water collection. As mentioned above, the request may originate from the agriculture or health ministry, who perceive particular benefits such as increased settlement of a fertile area, or reduction in a particular disease.

Technical consistency

It has already been mentioned that a rural water supply programme should be divided into sub-programmes if different technologies (e.g. piped, dug-well or borehole systems) are involved. Similarly each sub-programme should limit itself to reasonably uniform technical practices. For example, it is infinitely preferable to utilize sources that do not require treatment at all. Where such sources are not available, the programme may gradually develop a capacity to carry out minimal simple treatment. It would be wrong for a programme to undertake a wide range of technical practices, such as different and complex forms of treatment or pumping. A proposed project requiring such measures should be ruled out or passed to another implementing agency. In Malawi, the programme concentrated initially on gravity-fed piped systems from surface streams that did not require treatment. As the programme grows in strength a pilot project may be undertaken to construct, say, a slow-sand filter, or to construct a simple pumped scheme, but such developments should not be allowed to drain the manpower resources from the regular established programme.

Other criteria

Each programme will develop its own criteria, such as the maximum length of pipeline, or maximum population involved. For smaller projects, the presence of a school, health post or other development services may influence the selection process. Political considerations will also affect the final list. The government may wish to favour a particular area that has been neglected in the past, or show good will to an area which may pose a political problem.

3.6 Project Preparation

Procedure for project selection

Preliminary estimates of cost and manpower requirements have to be made for each application and then a short list of selected projects should be drawn up in accordance with the above guidelines. In Malawi projects were divided into major schemes suitable for submission to individual donor agencies and minor schemes which were financed by a special "rolling fund". The list was processed through the Minister for approval in order to minimize interference at a later date. The list was also forwarded to the central development authority, known as the Development Division, to ensure that it did not duplicate or clash with the activities of other ministries.

Project submissions

Whatever the source of funding, whether it be government itself or a donor agency, projects should be submitted in a standard form for approval. Major projects for individual funding should be submitted separately; smaller projects may be lumped together. The submissions should show the cost and disbursement schedule required, the aim of the project, completion date, background information, description of the work, division of responsibility, and financial details.

3.7 Supplies

Procurement is usually a function of the programme headquarters, although in some countries this may be done by a Ministry of Supply or its equivalent. Nevertheless it is important that procurement is closely monitored by programme staff to ensure that materials are procured in accordance with the specifications. Any necessary change in specifications should be referred back to the headquarters and, if necessary, to field engineers before procurement is authorized.

Standards and specifications should be uniform for the whole programme as far as possible, to facilitate interchangeability and to reduce the problem of storage.

The timing of procurement is crucial as it is essential that projects are not delayed through lack of supplies. Ideally a stock of materials should be built up at central or regional stores to "cushion" the effect of inevitable delays in delivery. In Malawi, major projects had their own separate stores on site to reduce the load on the central store.

Local procurement is obviously preferable to overseas procurement for reasons of flexibility, delivery, and cost, provided the quality is adequate. Even if the quality is substandard, it is in the interests of the programme to encourage local manufacturers to improve their quality by placing small orders and offering bigger ones conditional on, say, installing better equipment, or improving their quality control procedures.

CHAPTER 4
Field level organization and management

4.1 Introduction

The previous chapter dealt with the organization and management functions carried out at the central level. However, the real strength or weakness of a programme lies in its organization and management in the field. A programme with a weak central management may survive and even develop, as long as the field is strong. In Malawi the programme was originally much stronger in the field than it was at central level—the policies and management functions at the centre in fact developed out of the field experience. A major reason for this was that the head of the programme was himself working at the field level in the early years.

Technical supervision

In general, rural water supplies are technically simple compared with urban supplies. Because of this, some programmes have made the mistake of assuming that water systems can be constructed by the people with virtually no technical supervision. This assumption has the advantage that political pressure can be relieved merely by giving out pipes and materials to communities. Sometimes the system may actually be constructed, but because the *quality* of the work is so poor (e.g. pipes laid on the surface) the system invariably lasts only 1 or 2 years, and the political pressure builds up again. Adequate technical supervision at field level is one of the keys to a successful programme. The organization and management at field level has to be designed to provide that supervision.

Community participation

A second common mistake made by some programmes is that, as the people obviously want water, community participation will be *automatic*. The community is expected to manage its own contribution to the project, usually in the form of labour, while the government merely provides the materials. The reality is usually very different. If it is a large-scale project, several communities will be involved and will have to cooperate in a way that they probably have never done before. Even in a small project the community will have very little idea of the importance of certain technical requirements, such as the trench depth of a pipeline, and may never have worked as a community on a project involving virtually every individual for a substantial period of time.

For the community, the project is a "one-off" experience in which, if unassisted, they will inevitably make a number of mistakes and learn a number of lessons. The programme field staff, however, will have accumulated experience on previous projects and will know only too well the common pitfalls and mistakes to avoid. It is only natural therefore that the field staff must take an *active role* in advising and guiding the participation of the community.

The Malawi experience

It is at the field level that the experience of the programme in Malawi is particularly relevant. Although conditions vary from country to country, most of the activities necessary to construct a water supply system with the participation of the community are common. For this reason, the account below concentrates particularly on the experience of Malawi in order to present a "whole" picture of an organization that is actually working. The reader may extrapolate or interpret this account to suit his or her own situation.

4.2 Project Organization

Field staff

Figure 4.1 shows the staff organization of a major project involving 75,000 people. The role of the staff is to provide the technical supervision to ensure a good quality of work and to help organize and manage the participation of the people. The Project Manager is a professional engineer with overall responsibility for the project. He is assisted by Supervisors who are experienced in the social and technical skills required. They are responsible for the day-to-day supervision of Project Assistants and for maintaining the self-help labour programme. The Project Assistants are responsible for the routine management of self-help labour, the maintenance of technical standards, and the performance of certain technical tasks. Job descriptions for field staff are given in Appendix 2 and further aspects are discussed in Chapter 5.

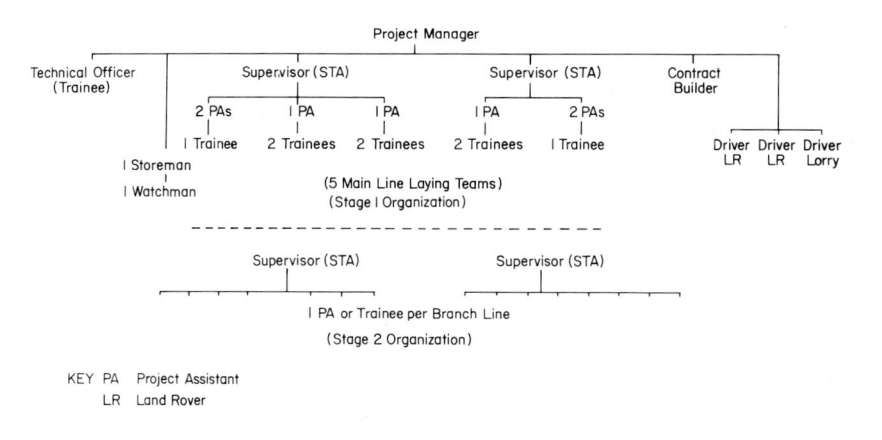

```
                              Project Manager
          ┌──────────────┬────────────┴───────────┬──────────────┐
  Technical Officer   Supervisor (STA)       Supervisor (STA)    Contract
     (Trainee)        ┌──────┬──────┐       ┌──────┬──────┐      Builder
        │           2 PAs   I PA   I PA    I PA   2 PAs            │
     I Trainee        │      │      │       │      │         ┌─────┼─────┐
        │         I Trainee 2    2      2      I Trainee  Driver Driver Driver
    I Storeman         Trainees Trainees Trainees                LR   LR  Lorry
        │
    I Watchman             (5 Main Line Laying Teams)
                             (Stage I Organization)
          ─ ─ ─ ─ ─ ─ ─ ─ ─ ─ ─ ─ ─ ─ ─ ─ ─ ─ ─ ─ ─
                   Supervisor (STA)        Supervisor (STA)
                   ┌───────┴───────┐       ┌───────┴───────┐
                     I PA or Trainee per Branch Line
                           (Stage 2 Organization)

         KEY  PA   Project Assistant
              LR   Land Rover
```

Figure 4.1 Staff organization for a major project in Malawi

Field level organization and management

```
Related Levels of
Project Organization
                              Chief + District Party Chairman
PM, STA                       Main Project Committee
STA, PA                       Section Committee                STAGE I ORGANIZATION
                                                               FOR MAIN LINES
PA                            Village Committees, Headmen, Leaders
PA                            Villagers
------------------------------------------------------------
PM, STA                       Main Project Committee
STA, PA                       Branch Committees                STAGE 2 ORGANIZATION
                                                               FOR BRANCH LINES
PA                            Village Headmen, Committees, Leaders
PA                            Villagers

        KEY  PM   Project Manager
             STA  Supervisor
             PA   Project Assistant
```

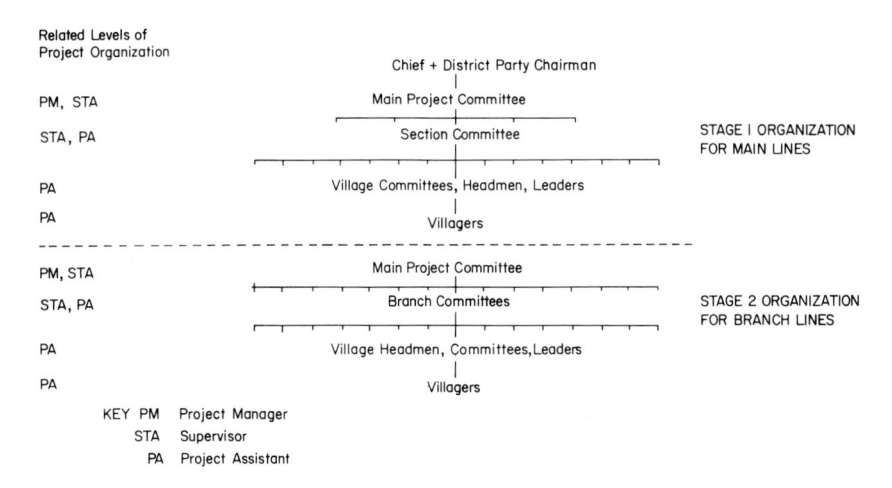

Figure 4.2   Community organization for a major project in Malawi

A project of this size is executed in two stages. The first stage involves the installation of the main pipelines, for which a team of two or three Project Assistants is allocated to each section of pipeline. The second stage involves the installation of branch pipelines, for which a Project Assistant is allocated to each branch.

Community organization

The community is organized under a number of committees at different levels. The role of the committees is to set up and maintain the self-help labour programme. The organization is quite distinct from that of the field staff, although naturally there is close cooperation between the two. Figure 4.2 shows how the community is organized for the two stages of a major project, and the relationships between project staff and the community structure. This relationship is not rigidly defined, and the overlap reflects the close collaboration between all levels of staff. The roles of the individual committees are discussed in Section 4.4.

4.3   Project Planning

Major projects

The smooth running and ultimate success of the project is highly dependent on careful planning and preparation. It is essential to avoid delays that could jeopardize the self-help commitment. Particular attention must be given to the timing of staff appointments, the delivery of pipes and vehicles, the transport of materials, and the motivation of committees. Exact timing and sequence varies from project to project, but certain activities are fixed by seasonal factors. These are:

(1) the construction of the intake before the rains;
(2) the marking of the pipeline before crops are planted;

Table 4.1 Schedule for a major project in Malawi
(The letters P and C indicate activities in which the Project Staff and the Community are involved respectively)

Year 1
1. Apr — Beginning of financial year, funds for new project become available (P)
2. Apr/May — Procurement of plant and vehicles (P)
3. — Staff transfers, at least sufficient to initiate project (P)
4. May/Jun/Jul — Site and construct project headquarters/store and any staff houses (P)
5. Jun/Aug — Public meetings to announce the Project (P,C)
6. — Forming main project committee to start initial programme (P,C)
7. Jun/Jul — Site intake, screening, and sedimentation tanks (P)
8. Aug/Sep — Construct intake (P,C)
9. Aug — Survey and site main line storage tanks and river crossings (P)
10. Aug/Sep/Oct — Mark main line from aerial photographs (P,C)
11. — Chain and label main line (P)
12. — Clear road along main line for asbestos cement (AC) pipe delivery (P,C)
13. Aug/Sep/Oct/Nov — Delivery of asbestos cement pipes from railhead to pipeline (P,C)
14. Oct/Nov — Survey main line (P)
15. — Construct river crossings if required (P)
16. Oct/Nov/Dec/Jan — Construct screening and sedimentation tanks (P)

Year 2
17. Feb — Formation of section committees to organize the main line trench digging programme (P,C)
18. Feb/Mar — Start trench digging programme (P,C)
19. May/Jun — Start asbestos cement pipelaying as soon as possible (P,C)
20. Jul/Aug — Completion of trench digging though some obstacles may still remain (P,C)
21. — PVC pipe arrival, delivery from railhead to pipe store (P,C)
22. — Excavate main line storage tanks (P,C)

---

(3) the start of the main digging programme in a period of relative lull in agricultural activity, between planting, weeding, and harvest (some special sections may be started earlier);
(4) the start of the main laying programme, the target for the year's work being calculated on what may be laid in the 5 or 6 months before the rains.

Other project activities should be planned around these dates. Table 4.1 shows a schedule for a major project of 3 years' duration, including 1 year of asbestos cement pipelaying. The table is also useful as a list of all the activities involved in the execution of a project.

Field level organization and management

23.		Stock building materials (P,C)
24.	Sep/Oct	Connection of main line sluice valves, air valves, flush points, and river crossings (P)
25.	Oct	All asbestos cement pipelaying sections link up and main line pressure-tested (P)
26.		Start construction of main storage tanks (as soon as water is "on") (P)
27.		Survey and site branch line storage tanks (P)
28.		Mark branch lines from aerial photographs (P,C)
29.	Oct/Nov	Inspect and protect main line from surface rain water damage; check after first heavy rain (P)
30.	Dec/Jan	Sow *Paspalum* grass seed to mark main pipelines (P,C)

Year 3

31.	Feb	Formation of branch committees to organize branch line digging programme (P,C)
32.	Feb/Mar	Start branch line digging programme (P,C)
33.	May/June	Start PVC pipelaying as soon as possible (P,C)
34.	Jul/Aug	Excavate branch line storage tank sites (P,C)
35.		Stock construction materials (P,C)
36.	Sep/Oct	Completion of pipelaying on major branches, though lesser branches may continue through rainy season (P,C)
37.		Start construction of branch line storage tanks (P)
38.	Oct/Nov	Inspect and protect major branches from surface rainwater damage; check after first heavy rain (P)
39.		Commence construction of standpipe aprons (P,C)
40.	Nov	Commence connection of village taps to standpipes (P)
41.	Dec/Jan	Sow *Paspalum* grass seed to mark branch lines (P,C)

Year 4

42.	Jan/Feb/Mar	Complete connection of taps (P,C)
43.		Set up maintenance procedures (P,C)
44.		Carry out project inspection (P)
45.	Mar	Project completion by end of financial year (P)

Minor projects

A smaller project is also affected by the seasonal considerations given above, but the smaller size of the task enables a greater degree of flexibility in the schedule. The main digging programme can start in March or April of the first year, but the survey work and siting of intakes and tanks can be done in the previous dry season. It is sometimes possible to work on tank sites and the marking of main lines in the year before the project officially starts, provided that the project has been approved. No funds may officially be spent, however, until the beginning of the financial year in

Table 4.2   Schedule for a minor project in Malawi
(The letters P and C indicate activities in which the Project Staff and the Community are involved respectively)

Preliminary work
1.	Sep	Survey possible intake sites and alignment of pipelines from headworks; profiles sent to Senior Water Engineer (P)
2.	Oct	Construct intake (P)
3.		Site main tanks (P)
4.		Prepare access roads to tanks (P,C)
5.		Excavate tank site (P,C)
6.		Stock construction materials (P,C)
7.	Nov	Mark main pipelines from aerial photograph (P,C)

Year 1
8.	Feb/Mar	Public meeting to form project committees (P,C)
9.		Organize trench digging programme (P,C)
10.	Mar/Apr/May/Jun	Trench digging programme (P,C)
11.		Tank construction programme (P)
12.	May/Jun/Jul/Aug	PVC pipelaying programme★ (P,C)
13.		Construction of river crossings (P)
14.	Aug/Sep/Oct	Construction of standpipe aprons (P,C)
15.		Connection of village taps to standpipes (P)
16.	Oct/Nov/Dec	Protect lines from surface rainwater damage (P)
17.		Carry out project inspection (P)
18.		Train committees in maintenance procedures (P,C)
19.	Dec	Sow *Paspalum* grass seed to mark pipelines (P,C)
20.		Project completion (P)

★ Assuming PVC pipes available from stockpile; otherwise pipelaying is delayed until pipes arrive in July or August.

April. An early start depends on the availability of project staff, and is facilitated by the proximity of the new project to a current one. Table 4.2 shows a schedule for a minor project of 1 year's duration.

4.4   Community Participation

The strength of any self-help labour programme is dependent on two factors: first, the way in which agreement is reached between the government and the community to cooperate on the project, and secondly, the establishment of proper authority for committees to carry out their work.

Initiating the project

The traditional and political leaders of the project area are consulted as to a suitable date for a public meeting to announce the project. If more than one area (e.g.

traditional or administrative) is involved, a separate meeting must be arranged for each. The meeting is conducted jointly by the traditional and the political leaders (e.g. MPs, District party officials) and attended by the head of the district administration, representatives of the District Council and of the Local Development Committee, and all project staff. Such a meeting is an important occasion, attended by several hundred people, and is usually preceded by national and political songs and dances.

As the meeting starts the local MP or, occasionally, District Administration official, announces that, in response to the request of the people, the Government has selected their area for a water project and is offering to provide pipes, materials, and technical supervision if the people are prepared to provide their labour. Usually the people have seen water projects completed in nearby areas, and everybody understands what this means. Nevertheless it is important that the traditional leader asks his people whether they accept the offer and that the people and the leader himself publicly agree to provide their labour. This procedure may seem an unnecessary formality, because the project would not have reached this stage of initiation unless the support of the community was already assured. But it is important that the Government and the people make their commitments publicly at the very start of the project, so that if and when problems arise later, the people concerned can be reminded of their commitment.

The project staff should then be introduced and the Project Manager asked to explain the work involved. He describes the source, the rough alignment of pipelines, the positions of tanks, the project boundary, and a rough schedule of the work. The Supervisor then explains what is required in terms of committees and the self-help labour programme. The election of committees is usually postponed to a later meeting of local leaders. These leaders are taken to visit the committees of a current project nearby, and a film describing a project is shown to the public at various centres in the area.

Authority and responsibility of committees

The self-help programme cannot be successful unless the committees are seen to have the *authority* to act. This authority *must* come from the local leadership, and cannot come from the project. Thus the committees are set up by the local leadership, whose authority is passed down from one committee to the next and finally to the people. With the traditional leader's authority behind them, the committees become *responsible* to him for their work, and when problems arise they feel obliged to resolve them. This responsibility is vital for the success of the self-help programme.

If the project staff were to try to set up committees, this authority would be absent. When problems arose the committees would not have the authority, in the eyes of the people, to act, and they would pass the problems to the project staff to solve.

The main committee

After the project has been announced the traditional leader calls a meeting to elect a Main Committee to be responsible for the overall management of the self-help programme. The first task of this committee is to organize the initial work programme. It is important that the tasks are well planned so that the project staff can make the requirements very clear. For example:

Task 1	Intake, trench digging from intake to sedimentation tank, excavation of screening, and sedimentation tank sites.
Requirement	Daily work programme involving villages living nearest.
Task 2	Marking the main line.
Requirement	Twenty men daily for about 2 weeks.
Task 3	Clearing access road along pipeline.
Requirement	Daily work programme involving villages living nearest to main line for about 2 or 3 weeks.

When these have been explained it is left to the committee to decide how the requirements should be fulfilled. For example, it may delegate the organization for the headworks to one committee member from that area, who may form his own committee of the villages involved. For marking the line and clearing the road the committee may contact the villages along the line and organize a work programme. Project staff will visit the relevant villages the day before they are required to give the final detailed instructions.

Once the initial work programme is completed, the Main Committee is responsible for setting up the main line trench digging programme (item 17 in Table 4.1). This is the first major task in which *all* villages are involved. About a month before the programme is due to start the Main Committee meets to discuss the work. The Project Manager explains the work to be done, and can suggest ways in which the work may be divided up. He is in a good position to advise on this as he has an overall picture of the project, and the location and population of all the villages involved. Nevertheless, it is left to the meeting to decide how the work should be divided, and once this decision is made the committee advises the villages concerned in each section to form their own Section Committees.

Section committees

Villages in each section then meet to elect the Section Committee, whose immediate task is to draw up a daily programme of villages to work on the trench. It is important that the committee is provided with a list of villages and populations so that a balanced work programme can be made. The committee decides how the task will be tackled, though project staff usually advise digging to commence at the upstream end of the section and work progressively down; sometimes the committee decides to divide the section up into subsections for each village. The committee produces a rota of committee members to attend the trench each day to help supervise the work. In addition, the committee asks each village to nominate four leaders to supervise the work of the village on its day of attendance.

The secretary of the Section Committee is provided with special letter-forms to remind each village the day before it is due for work. The secretary often also keeps a record of village attendance; any village showing consistently poor attendance is given widespread publicity so that social pressure is brought to bear on that village to improve its performance. The committee is therefore active, though it rarely meets formally.

Field level organization and management

If the section's digging programme gets off to a good start, it usually settles down to a steady rate of progress. With a long and arduous section, attendance at the trench usually begins to drop after 2 or 3 months. This may be attributed both to the harvest period and to the wearing off of the initial novelty and enthusiasm.

By this stage about three-quarters of the section may be completed, and it should be possible to start laying soon. The chairman of the committee should call a meeting to discuss the problems and remotivate the programme. It may decide to reorganize the work so that each village is given a section remaining to be dug, and it may detail certain villages to help the laying programme get under way on a completed part of the trench.

If attendance continues to be unsatisfactory, the committee can refer the matter to the Main Project Committee which may in turn ask the traditional or party leaders for help. A public meeting may then be called, after which attendance usually improves sufficiently to complete the work. Once the section is dug, the same work programme continues, but with the new task of pipelaying and backfilling as described in Section 4.5.

The installation of the main line represents the end of Stage 1 of the project. The Main Committee meets again to call upon the villages of each branch line (defined by the design) to form Branch Committees.

Branch Committees

The role of the Branch Committee is very similar to that of the Section Committee. The branch lines (usually laid in plastic pipe) are easier to dig as the trench depth is less than for asbestos cement pipes, and the villagers generally have less far to walk to work. In addition there is usually the added incentive that water pressure is already on in the main line.

The Branch Committee continues to function after the project is completed, with particular responsibilities for maintenance, cleaning of tanks, minor repairs, and checking abuse of water (see Chapter 7).

Village Committees

The leadership of the village may be entirely in the hands of the village headman, or it may also involve village leaders and, sometimes, a Village Committee. The village leadership is responsible for supervising the village labour on its appointed day of work, and for ensuring that village attendance is maintained. It is also responsible for selecting the sites for all standpipes allocated to the village by the project. After project completion, the village leadership is involved in maintenance as described in Chapter 7.

4.5 Management of Self-help Labour

Trench digging

A large project may require the digging of some 150 km of trenches in 6 months. This can only be achieved by efficient utilization of a large labour force. The

Figure 4.3 Villagers excavating a rock section of trench near an intake—men do most of the difficult work

(Photo: the author)

Figure 4.4 Rock-breaking-Project Assistants pour water to crack a rock which has been heated for 6 hours. This method was commonly used in Malawi

(Photo: A. Glennie)

Field level organization and management

Figure 4.5 A Supervisor attacking the rock with a hammer—in general villagers only work in the mornings; project staff work in the afternoons on jobs that do not require village labour

(Photo: A. Glennie)

Figure 4.6 Women do most of the easier work—procedures were later improved to heap soil on one side only

(Photo: the author)

methods are described here in some detail to show that self-help is possible on this scale if sufficient attention is paid to detailed organization.

In the first days of the programme village attendance is likely to be abnormally large. It is therefore essential that adequate preparations are made to ensure a smooth start. If there is chaos and confusion at the start, little progress will be made and the people will feel their journey has been wasted.

Preparations are made on the day before work is due to start. The first 300 m of the line is cleared and marked to avoid hundreds of villagers having to wait around the following morning while this is being done. Tools, marking string, and measuring sticks are all prepared beforehand, a measuring stick being placed every 20 m along the line. (This is a piece of bamboo cut to the required depth of the trench.)

Once the digging programme starts it is essential that adequate organization is provided for all the necessary activities. The route will have been marked some months before, and although crops are not planted on the line, it is necessary to clear

Figure 4.7 The same scene as Figure 4.6 looking back towards the source. Mulanje Mountain has become the source of piped water for over 350,000 people living on the plain—all installed by self-help

(Photo: the author)

Field level organization and management 49

Figure 4.8 Men and women work together in heavy clay soil—committee member on extreme right

(Photo: the author)

grass and bush again before digging. One leader is therefore sent ahead with about 25 villagers to clear the line for the next day's work. Another leader and about 10 men are taught how to mark the trench outline on the ground using the string provided, and to divide the trench into 3 m sections. They do not mark across gullies and streams as these are not dug until the rains stop. This is to ensure all surface water is able to cross the line and does not break into the trench where it would cause serious damage. These sections are eventually dug immediately before laying.

A third leader, preferably a committee member, is selected to allocate each villager to a section as he or she arrives for work. A common allocation is one man and one woman for each 3 m section of ordinary clay soil.

All village leaders are given a measuring stick to ensure each section is dug to full depth before the villagers are permitted to leave. Great emphasis is laid on this procedure so as to establish the routine from the very beginning. The standard is then much easier to uphold throughout the project.

The project provides a selection of picks, shovels, crowbars, axes, and 7 kg hammers which are distributed by the Project Assistants. Villagers are expected to bring their own hoes for digging.

The Project Assistants assist the leaders generally in their supervision and concentrate particularly where problems are encountered, such as on rocky or laterite sections. After the day's digging is over, usually by 11 a.m., they inspect the trench to check the depth and width so that a few people may be allocated the next day to clear up any minor problems. Major problems may take several days to

Figure 4.9 A team of villagers under the supervision of a village leader carry asbestos cement pipes along the trench ready for laying

(Photo: L.H. Robertson)

Figure 4.10 Villagers carrying asbestos cement pipes

(Photo: L.H. Robertson)

Field level organization and management

Figure 4.11 Project Assistants join asbestos cement pipes in the afternoon after villagers have laid the pipes down in the trench. The Supervisor is on the extreme right (Photo: L.H. Robertson)

overcome; it is preferable to overcome these at the time rather than to leave them to be tackled later.

Laying of asbestos cement pipe

Asbestos cement pipelaying is carried out on each section by a team of two or three Project Assistants with the assistance of self-help labour. High laying standards can be achieved under close and sustained supervision by Supervisors and the Project Manager. To ensure maximum supervision laying is started on one section at a time with a Supervisor working with each team for its first week.

Efficient laying is dependent on organization and preparation. The task of the self-help labour is to carry out all the preparatory unskilled work:

(1) As the trench is usually dug some months before, one leader and about 10 people clear out the earth and debris that may have fallen in.
(2) One Project Assistant and about six people backfill about 75 mm of soft soil into the trench and rake out the bed to a smooth level surface.

(3) A third group carry pipes from the dumps and lower them into the levelled trench end to end. (Asbestos cement pipes are delivered by lorry to dumps every 100 or 200 m along the line, usually before the trench is dug. A temporary access road is cleared by self-help for this purpose.)
(4) A fourth group fit the rubber O-rings into the collars, smear them with lubricating soap, and place one collar beside each joint.

These preparations are continued until the villagers leave for home in the late morning. The Project Assistants stay behind to join all the pipes prepared. With this system about 100 pipes can be joined every day, covering a distance of 400 m.

Backfilling

All villagers not engaged in the preparation of pipes are organized into groups for backfilling. This is one of the most important parts of pipeline installation but there is a danger that it receives the least supervision, as it appears to be the "easy" part of the operation. To avoid large numbers of villagers enthusiastically but haphazardly heaping the soil back into the trench, the backfilling is divided into three stages:

(1) The first stage is the most important, and is therefore entrusted to a responsible leader with about 10 men. Relatively smooth soft soil is backfilled just sufficient to cover the pipe and then compacted by pounding with feet.
(2) A second group follow at a distance of about 20 m. A leader and about 20 people backfill the trench to half-depth and compact again.

Figure 4.12 Backfilling is a popular activity which always needs careful management and supervision. Project Assistant in foreground   (Photo: L.H. Robertson)

Field level organization and management

(3) The final stage is carried out by the rest of the villagers. All the remaining soil is backfilled and built up into a ridge over the pipeline. The ridge is discontinued across gullies and waterways which must be left free for rainwater to cross unhindered.

It was found through experience that backfilling should be completed at the same time as the pipelaying programme. Originally the practice was to backfill half the trench and leave the collars exposed until the pipeline had been pressure-tested. But problems were experienced in remotivating the villagers weeks or months later to complete the backfilling at a time when they were usually busy digging their branch lines. With plenty of labour it is relatively easy to dig up the few burst pipes that may be exposed during testing.

Laying of PVC pipe

The preparations for PVC pipelaying are similar to those for asbestos cement pipes, but the flexibility and robustness of PVC allows a greater margin for error. One group clears the trench ahead, while another carries the pipes out and lays them alongside the trench. The pipes are joined beside the trench and the solvent cement joints are allowed to stand for 30 minutes before the pipeline is laid into the trench. Backfilling is completed in one operation, but is not carried out after 9 a.m. to avoid warm pipes being backfilled in an extended state (see Chapter 6).

Figure 4.13 PVC pipes can be joined by villagers under adequate supervision

(Photo: the author)

Figure 4.14 Villagers are brought by project transport to unload asbestos cement pipes at the nearest railhead—breakages were very rare

(Photo: the author)

Figure 4.15 PVC pipes being unloaded and stacked neatly under the supervision of Project Assistants (Photo: the author)

Field level organization and management 55

Figure 4.16 PVC pipes being unloaded at a village under supervision—note nesting of pipes to save volume

(Photo: the author)

Figure 4.17 Villagers extracting PVC pipes from a pipe store at a project headquarters. PVC pipes should be stored in the shade for any significant period of time

(Photo: H. Van Schaik)

Other self-help activities

Other self-help activities that are managed by Project Assistants are: the marking of pipelines from aerial photographs; the clearing of temporary access roads to tanks and along main pipelines; the excavation of tank sites and the collection of river sand for tank construction; the loading and unloading of asbestos cement and PVC pipes from the nearest station to the project; the planting of *Paspalum* grass seed to mark completed pipelines; and the construction of village standpipe aprons.

4.6 Management of Project Staff

The degree of responsibility given to project staff is a highly significant feature of the programme in Malawi. A team of three Project Assistants, for example, may be

Table 4.3 List of contents of field handbook for Project Assistants in Malawi

		Pages
1.	Organization and Committees	1– 8
2.	Trench Digging	9
3.	Installation of Asbestos Cement Main Lines	10–17
4.	Procedure for Laying Asbestos Cement Pipe	18
5.	Instructions for Asbestos Cement Joints and Fittings	19–21
6.	Installation of PVC Lines	22–23
7.	Procedure for Laying PVC Pipes	24
8.	Instructions for PVC Pipe Joints	25
9.	Making Heads on PVC Pipes	26
10.	Procedure for Laying Steel Pipes	27
11.	Village Tap Sites	28–29
12.	Maintenance Duties for Completed Projects	30

Forms

F1	Ntchito ya Madzi★
F2	Kalata Kwa Atsogoleri Onse†
F3	Fortnightly Work Programme
F4	Weekly Report (Progress)
F5	Village Tap Check Sheet
F6	Village Tank Check Sheet
F7	Stores Chit
F8	Tank Inspection Report
F9	Village Tap Inspection Report
F10	Mainline Maintenance Inspection Report
F11	Branchline Maintenance Inspection Report
F12	Builders Check Sheet—Tank Construction
F13	Main Tank Maintenance Inspection Report
F14	Weekly Report (Maintenance)

★ Letter form for the use of Committee Secretaries calling Village Headmen for work.
† Letter concerning maintenance read out and distributed to leaders at the tap installation ceremony.

Field level organization and management

responsible for an 8 km section of main asbestos cement pipeline involving 15,000 people; or a single Project Assistant may be responsible for 25 km of branch lines involving 5,000 people. The selection, training, and motivation of these Project Assistants play an important part in preparing them for this responsibility (see Chapter 5). Once in the field, however, the Project Assistants need to be supported within a framework of standard procedures, regular visits from Supervisors, and regular staff meetings.

Standard procedures

In the Pilot Phase of the programme, before the development of routine procedures, project activities were carried out on an *ad hoc* basis. The recruitment of the first Project Assistants in 1972 led to the need for standard procedures to facilitate supervision and ensure the raising of technical standards. Most of the procedures evolved from the experience gained on the early Mulanje West and Phalombe "consolidation" projects, and are now laid down in a handbook for Project Assistants. The contents of this handbook, which is the manual used for initial training, is given in Table 4.3.

Technical operations, such as the making of joints in asbestos cement, PVC or steel pipes, or the construction of standpipe aprons, are spelt out in step-by-step detail. Procedures for the management of activities involving self-help labour, including trench digging and pipelaying as described in Section 4.5 above, are also laid down in detail. Forms and check-lists are provided to assist Project Assistants with special activities.

Fortnightly work programme

Project Assistants are required to submit a work programme to their Supervisor every fortnight. This is particularly relevant during the installation of branch lines when several activities are running concurrently. There may be four or five lines being dug, standpipe aprons to be constructed, and other lines where the Project Assistant's presence is required for pipelaying. The purpose of the fortnightly work programme is to encourage him to plan out his work as efficiently as possible, and so that he can give sufficient notice to the various villages of the day that he intends to visit them. The work programme is also part of the process whereby the Project Assistant is made to feel responsible for his work. However, he is not expected to adhere rigidly to his programme as circumstances inevitably change. The programme also lets the Supervisor know where his staff may be working on any day.

Weekly report

Project Assistants are also required to submit a report form every week. Again, this is of particular relevance during the installation of branch lines. The purpose is to record progress of trench digging, pipelaying, apron construction and tap connections, as well as events such as committee meetings and pipe bursts. It is possible that this report is superfluous as it duplicates information that is recorded on master charts during the weekly staff meeting. However, the discipline of having to write a report is a contributory factor to the overall motivation of Project Assistants, as long as the report is strictly factual and easily completed. The form covers one side of foolscap paper and can be completed in about 15 minutes.

Weekly staff meeting

The weekly meeting of all project staff is an event of major importance in the work programme, and deserves more detailed description. The meeting is held at project headquarters during the afternoon, after the morning's work. Before the meeting starts, Project Assistants write out and submit their work programmes and reports to their Supervisor, and take the opportunity to draw from the store whatever fittings and materials are needed for the coming week. Each Project Assistant in turn makes a verbal report to the Project Manager who plots the progress of work on the master aerial photograph mosaic displayed in the headquarters. The Project Manager also brings up to date the progress charts for the tank construction and transport programmes.

After all this preliminary activity, the meeting itself starts with each Project Assistant giving a very brief progress report. This gives all Project Assistants the opportunity to hear how their fellows are progressing, and is an important factor in the maintenance of the team spirit. The Project Manager and Supervisors raise points concerning the work they have observed that week, and certain procedures may be re-emphasized and practised, particularly for the benefit of trainee Project Assistants. The coming week's programme is discussed, particularly the transport programme which is arranged by the Supervisors in response to the requirements of the Project Assistants.

The meeting is also an opportunity for the raising of general administrative points and the discussion of complaints and suggestions from Project Assistants.

4.7 Tank Construction

The contract builder system

Experience gained during the Pilot Phase of the programme showed that the construction of reinforced concrete tanks was generally beyond the capabilities of village level artisans. This experience led to the decision to employ a local builder on a contract basis, and to give him and his team sufficient training and technical support to carry out all construction work, with materials supplied by the project.

This system has developed to the point where, on one project, the contract builder employed four construction teams to work simultaneously on different tanks, enabling the rate of tank construction to keep up with other project work.

For projects in a new area, a suitable local builder is found and sent to work with a construction team on a current project which is building a tank of similar size. He then returns to his own project to build an identical tank under close supervision.

This system has proved to be ideal for the construction of tanks of adequate standard at minimum cost. In 1976, the material and labour cost for a 225 m^3 (50,000 gallon) reinforced concrete tank was US$4,600, of which US$850 was paid to the contract builder. An urban-based civil engineering contractor would find it much

Field level organization and management

more expensive to construct relatively small works scattered throughout the rural area. In any case, the use of a local rural builder is preferable as it contributes both to rural development and to the rural economy. The direct labour system is an alternative but it would increase the administrative and supervisory load on the project, and could result in a lower rate of construction due to the lack of incentive compared with the contract system.

Procedure for tank construction

The tank site is chosen by the Project Manager according to the elevation required by the design, the ease of access, and ease of excavation. The site is then excavated by self-help labour, to a depth at least sufficient for half the tank to be buried below original ground level (which should leave sufficient backfill to cover the tank). An access road is also cleared so that heavy transport can reach the site. River sand and foundation stone are collected by self-help and delivered to the site in advance, along with stone aggregate and reinforcing bars.

Two members of the construction team are sent in advance to build up the foundation with rough stone to the level of markers set by level. The rest of the team are then transported to the site with all their equipment, which includes a concrete mixer and vibrator, shuttering, tools, and temporary sheds for accommodation and storage.

The builders are invariably illiterate, and certainly cannot read a construction plan. However, as a team specializing in the construction of tanks, they have no difficulty in remembering the stages of construction and the principal dimensions.

Figure 4.18   Villagers excavating a 50,000-gallon tank site

(Photo: the author)

Figure 4.19 A 50,000-gallon reinforced concrete tank under construction—the contractor could not read a plan but was trained by the project to construct tanks of all sizes

(Photo: the author)

Figure 4.20 The roof under construction—the contractor employed his own labour and was paid a standard rate for each tank size

(Photo: L.H. Robertson)

Field level organization and management
61

Figure 4.21 Finishing touches go on the roof—note the wall poured in rings

(Photo: the author)

Figure 4.22 A Project Assistant inspects a 600-gallon village storage tank built by the villagers. Villagers had to build their own tanks if they wanted more taps than their allocation

(Photo: the author)

For simplicity, the thicknesses of the floor, wall, and roof, and the height of the wall and central roof support pillar are standardized for all tank sizes. This means the smaller tanks are overdesigned, but the extra cost is negligible compared with the saving in supervision and the reduction of costly mistakes. The only information the builders need to be given is the diameter of the tank and the distribution of reinforcing bars in the wall. For this purpose, the contract builder is given a Tank Construction Check Sheet which indicates the distribution of reinforcing bars in diagrammatic form. Other information on the form can be read by the Project Assistant, who is also responsible for supplying the builder with the right pipes for casting in the walls of the tank (inlet, outlet, overflow, and drain).

The construction itself proceeds under the supervision of the contract builder with regular visits from the Project Manager (or Technical Officer in the case of a smaller project). The Project Manager is responsible for planning the tank construction programme to tie in with the laying of pipelines, so that water is available at the tank site when required. Tank construction can proceed throughout the wet season, but transport of materials and the digging of river sand is ideally scheduled for the dry season.

4.8 Transport

Table 4.4 shows the transport requirements of a major project for about 75,000 people.

Project Assistants are expected to have their own *bicycles* for which they receive a monthly allowance for maintenance. Those without a bicycle are encouraged to buy one, with a government loan repayable over 2 years. This system avoids the problems inherent in maintaining a fleet of project bicycles.

Supervisors and Technical Officers are issued with *motor-cycles* (*ca.* 100 cc) which are maintained and operated with project funds. The motor cycle is probably the most efficient means of personnel transport in rural areas, especially if long distances have to be covered daily along village paths and pipelines. A bicycle is inefficient as the Supervisor's travelling time would be excessive in relation to his supervision time. On the other hand a four-wheeled vehicle is inappropriate as it is limited to roads, tracks, and bridges, and is very expensive.

Project Managers have the use of a *1-ton pick-up*. This vehicle is suitable for the relatively long distances that a Project Manager has to travel, both within a large project area and between projects. The load-carrying capacity enables stores to be delivered at the same time as supervisory visits are made.

Four-wheel-drive vehicles are used for the distribution of stores within the project area, principally PVC and steel pipes carried on an overhead frame, and construction materials for village standpipe aprons. The day-to-day management of these vehicles is in the hands of the Supervisors, while the Project Manager is responsible for overall operation and maintenance. These vehicles are not used as personnel carriers except when a large group of people needs to be transported.

Heavy transport is used for the delivery of asbestos cement and PVC pipes and tank construction materials. A 3-ton lorry is most suited to rural areas as it can relatively easily be pushed out of trouble, but transport costs are high per unit load and 5-ton

Field level organization and management

Table 4.4 Transport requirements for a major project in Malawi

Vehicle	Quantity	User/Controller	Functions
Bicycle	15	Project Assistants	Personal transport (personally owned)
Motor cycle	3	2 Supervisors 1 Technical Officer (Trainee)	Personal transport for field supervision
1-ton pick-up	1	Project Manager	Personal transport for field supervision; stores collection and delivery
4-Wheel drive pick-up with overhead frame	2	Supervisors	Distribution of PVC and steel pipes; distribution of construction materials for standpipe aprons; distribution of miscellaneous tools and materials as required
7-ton tipper	1	Project Manager★	Delivery of stone aggregate, sand, cement for tank construction
5-ton flat lorry	1	Project Manager★	Delivery of asbestos cement pipes, cement, reinforcing rod; transport of tank construction teams and equipment
7-ton flat lorry with high-sided superstructure	1	Project Manager★	Bulk delivery of PVC pipes

★ These three lorries would be shared with other projects in the programme on a rota basis controlled by programme headquarters.

and 7-ton vehicles are preferable where longer hauls are required. Larger vehicles are generally inappropriate for village tracks. Heavy transport comes under the control of the Project Manager and is usually based at a major project headquarters. Vehicles are, however, shared between projects, especially when there is a peak of transport activity, such as for asbestos cement pipe deliveries.

4.9 Project Stores Organization

Pipes and fittings

Asbestos cement pipes are delivered by the manufacturer according to an agreed schedule convenient for both the production and distribution processes. They are delivered by rail in monthly consignments over a period of about 4 months and transported *directly* to dumps marked along the pipeline, to reduce handling. Self-help labour is used for all loading and unloading of these easily breakable pipes, and it is remarkable that damage due to careless handling is almost nil. Although the pipes may remain at the dumps for several months, they are regarded as community property and are left alone. Accessories and fittings, however, require closer control and are stored at the project headquarters until required.

PVC pipes for major projects arrive in one consignment which is delivered straight to a shaded store prepared at project headquarters. The pipes for a number of smaller projects are ordered together, and arrive as one consignment. These are transported to the central programme store for later distribution.

Other materials

Tank construction materials are delivered direct to site from the supplier. All other stores are delivered to the project headquarters and reissued from there. These stores are usually purchased as required, though a small stock of the most common items (cement, plumbing materials, etc.) is maintained at project headquarters.

Stores requisitioning

A simple chit system is used for Project Assistants to requisition stores from the project headquarters. Most fittings are freely available for Project Assistants to help themselves, but the more attractive items (taps, cement) are issued personally by a Supervisor. Stocks are maintained by the Project Manager who either requisitions from the central programme store or purchases direct from the nearest supplier.

CHAPTER 5
Field staff

5.1 Introduction

The performance of field staff is the one factor that can be singled out as the most crucial to the success of any rural development programme. No matter how well a programme is planned, if the field staff are not well motivated the programme *cannot* succeed. Lack of motivation results in the inability to develop and to *overcome* problems. Motivation is not just a question of a decent salary. It is the result of a host of factors, including selection, training, leadership, responsibility, and comradeship.

Undoubtedly one of the most striking features of the programme in Malawi is the high performance of its field staff who, with relatively little training, assume enormous responsibility and enjoy the confidence and respect of the communities they serve. This chapter examines these issues in detail. The principal features are applicable to most rural development programmes anywhere in the world.

5.2 Policy

All government personnel have to fit somehow into the government's standard staffing structure. Generally, this structure is fairly rigid and it is not possible to change rules and regulations. Unfortunately, very often these rules are more geared to a formal, office-based, bureaucratic structure than they are to the needs of development. A development programme may need a large number of field workers who can be trained in a short, specific training course, yet government regulations may stipulate all field workers must have a certain minimum educational qualification and a further 2 years training; the result is that there are hardly any eligible people for the posts. Nearly all developing countries face this shortage of trained technical manpower. While one of the solutions is obviously to train more people, another is to rationalize manpower policies so that necessary staff can be given short, specific training in a particular programme and relatively large numbers of staff turned out. It is better to have larger numbers of less highly trained staff than a few highly trained people, because the most important factor in staff performance in rural development is on-the-job *experience* gained under adequate supervision.

In Malawi, staffing policy developed as a result of the needs of the programme. Initially, from 1968 to 1971, projects were staffed by regular government workers transferred from other duties. As mentioned in Chapter 2, the lessons learnt from the Pilot Phase included the requirement for a closer level of field supervision, implying a rapid increase in the number of field technicians, and a more specific and appropriate level of training. The lowest level of regular government staff were the Community Development Assistants who were used in the Pilot Phase, but they were very few in number and had to undergo a 2-year academic training which was inappropriate to the programme's needs. Some alternative clearly had to be found. Fortunately, the Ministry of Agriculture had faced this problem before, and the Government had specifically created a category of staff who were appointed only for the duration of a project. Their salary and allowances were on the same scale as permanent Civil Servants. Other countries have tackled this problem in a similar way. It is only a temporary solution, as invariably the staff have no career structure or promotion prospects. This temporary solution was adopted in Malawi, and a level of staff called Project Assistant was created. Twenty Projects Assistants were recruited for the two major "consolidation" projects in 1972. As the programme expanded, experienced Project Assistants were transferred to other projects and new staff were recruited for the major projects. The major projects were used as a training ground for the whole programme. The number of Project Assistants recruited were 20 (1972), 10 (1974), 20 (1976), 10 (1977), and 10 (1979).

5.3 Selection Criteria for Field Technicians

The selection criteria used in Malawi are applicable to most rural development programmes that need relatively large numbers of field staff. The term 'field technician' is used below, as this is a more generally applicable description than the title 'Project Assistant'.

Age

Applicants should be at least 25 years old. Field technicians are expected to communicate with community leaders who are usually senior and respected citizens. A person less than about 25 years of age is often regarded as a youth in many traditional societies and would not be able to gain the respect and confidence of leaders or villagers. The upper age limit should be about 40 years.

Education

Great care must be taken to choose the right educational level so that field staff are capable of being trained to do the job, yet are not unnecessarily highly educated. In most developing countries, very few people attend secondary school; the vast majority either have attended primary school or have had no schooling at all. The few secondary school leavers naturally consider themselves to have risen 'above' their fellow villagers and are often unable to reconcile their education and aspirations with the rural environment. They tend to 'look down' on rural people as uneducated, backward, and uncooperative. These people aspire to 'white-collar' jobs in the town and regard a job in a rural area as second-best. They are also generally "allergic" to manual labour and practical work.

It is essential that field technicians are socially compatible with rural communities, and their technical skills must be practical rather than academic. It is, therefore, preferable to recruit staff with *primary school* education. There are exceptions when a secondary school leaver may make a first-class field worker, and may show considerable potential for eventual promotion to the Supervisor category, but these exceptions should only be made if the candidate is personally known to one of the selectors, or if he has outstanding and reliable recommendations.

Staff members should be able to speak, read, and write in the national language.

Previous employment

In general, preference should be given to applicants who have had some form of satisfactory previous employment, as such people will be more able to bear the considerable responsibility given to project field staff. Applicants displaying outstanding qualities, however, should be considered even without previous employment.

5.4 Recruitment Procedure

Timing

The recruitment and selection process should be carried out in good time so that successful candidates can be appointed at the beginning of the work season. From the time of advertising, at least 1 month should be allowed for the receipt of application forms, a further 2 weeks before interviews take place, and the selection/training course should begin at least 2 weeks after the interview. The intervals will be longer in areas of poor communication.

Applications

Applications should be sought from the districts in which the programme is active. Vacancies should be advertised by notices displayed at the district level government offices, post offices, offices of the traditional leaders, markets, bus stations, and any other places where villagers regularly meet. Application forms should also be made available at these points, and applicants should forward the completed forms to the programme office, giving basic information including age, education, and previous work. These forms should then be sorted according to the criteria above and a reasonable number (six times the number of vacancies) should be called for interview.

Interview

The interview board should consist of the engineer in charge of the programme in that area, one experienced Supervisor, one representative from ministry headquarters if possible, and one 'outsider', such as a district officer of another ministry. The interviews should be held in the district concerned.

As candidates arrive at the place of interview they may be given a simple written intelligence test of the multiple-choice type, lasting about 30 minutes. This test need not be used to grade candidates, but it may affect the decision on borderline cases.

The purpose of the interview itself is to form an initial impression as to whether the candidate is the right material to *proceed to the second stage* of the selection process. In order to interview as many candidates as possible, each interview need last only 5 minutes, including 1 minute for awarding marks. The candidate should be encouraged to talk about himself, his family, and his previous work so that he is given an opportunity to portray his personality to the board. Appropriate personal qualities are considered to be most important, for experience has shown that the extrovert, smart, and articulate candidate does not necessarily make a good field worker. The attitude of the candidate to rural life is particularly important.

After the interview, the board should discuss the candidate briefly and each member should give him a mark. This method relies heavily on the judgement and experience of the interview board, but it has worked well in Malawi. When all the interviews are completed, the candidates should be placed in order of merit according to marks awarded, and the appropriate number (if possible, twice the number of vacancies) should then be called for the selection/training course.

5.5 Selection/Training Course

Purpose of the course

This is the most important stage of the selection procedure. Normally, government selection procedures finish with the interview, after which selected candidates are appointed and sent for training. However, experience has shown that, especially for this level of personnel, performance at an interview can be very misleading, and success in rural development depends so heavily on the quality of field staff that an interview alone is not sufficiently rigorous to eliminate unsuitable candidates. By contrast, a candidate's true qualities can be revealed remarkably quickly during an intensive course.

The purposes of this course can be summarized as follows: first, to give all candidates the time and opportunity to display their true qualities; secondly, to give the selecting staff time and opportunity to get to know each candidate; thirdly, to teach the technical skills and organizational procedures required of field technicians; fourthly, to impart a spirit of motivation.

Site of the course

Field technicians live and work in the rural areas, sometimes under difficult conditions and in temporary accommodation. It would be inappropriate and misleading, therefore, to run the course in one of the formal government training institutions, which generally follow an urban-oriented, academic pattern; it is more realistic and appropriate for the course to take place in a rural area similar to where the work is to be carried out. A certain amount of privacy and isolation is nevertheless desirable to avoid the distractions and constant scrutiny of village life.

A camp-site or local institution in or near a project area is ideal for the purpose, as it has the additional advantage that the resources of the project are available for the benefit of the course.

Duration and numbers

The duration of the course should be as short as possible. The purpose of this course is to select the right candidates, not to impart *all* the training necessary to carry out the job. The main part of the training will take place on the job itself, during the first 2 years of appointment. Courses in Malawi lasted from 2 to 3 weeks. This is the absolute minimum time necessary for the selection of good candidates, and it is only possible because of the very close supervision and high level of on-the-job training that trainees receive in their first 2 years. In other programmes this may not be possible, and so a longer course will be necessary to impart a higher level of knowledge and skills, so that trainees can manage with less supervision. The maximum duration of such a course should be 6 weeks.

The ideal number of candidates for a 2-week course is about 20. On the expectation that only half the candidates will be selected, it will be necessary to run more than one course if there are more than 10 vacancies. This may be impractical if there is a large number of vacancies and if it is necessary to run a longer course. In such cases, the absolute maximum is 40 candidates for a 6-week course.

Training staff

The course should be run by experienced programme staff. The best instructors are *experienced* Supervisors and field technicians as their experience is generally more relevant than that of the engineers, and they are usually able to convey their experience in a more colourful way. Ideally there should be a maximum of six or seven candidates per instructor. It is *not* advisable to leave the training to staff of a training institution who have not had experience of the programme. As the programme develops and expands the training needs will become so great that it will be necessary to appoint a full-time Training Officer who may be an experienced engineer or, more likely, a Supervisor. Continuity is important so that experience in conducting courses is accumulated. This Training Officer should also be responsible for in-service training courses, refresher courses, and on-the-job training.

Description of the course

The nature of the course will vary from programme to programme. To give an idea of one type, the course in Malawi is described here.

The candidates are divided into four groups for the duration of the course, and are issued with name badges, and some stationery. They are given 2 days to elect a camp leader who is responsible for domestic arrangements, including buying food at the local market. A local villager is employed as a cook.

The daily programme is divided into three sessions (see Table 5.1). During the morning sessions, from 6 a.m. to 9 a.m., each group progressively digs a length of trench, specially selected to give the candidates training in the use of all tools at their disposal. The purpose of this session is also to make the point that field technicians are expected to join in physical labour, and cannot supervise village labour unless they are prepared to tackle the problems themselves. In addition, the experience of working in a small group on a specific task, with an element of competition, helps to develop that team spirit which is a hallmark of the Malawi programme. During the

Table 5.1 Timetable for programme selection course in Malawi, 1979
(Arrival on Saturday preceding the beginning of course for preparation of camp site)

Week 1	6.00–9.00	9.30–12.00	13.00–15.00	Evening	Week 2	6.00–9.00	9.30–12.00	13.00–15.00	Evening
Monday	Trench digging AC*	Procedure for AC pipes	Reading maps and marking	Film	Monday	Protection of pipelines	Procedure for PVC	Organization and committees	Role games
Tuesday	Trench digging AC	Laying AC pipes	Aerial photographs organization		Tuesday	Preparing tap sites	Forming PVC heads	Gully crossings and checkdams	
Wednesday	Trench digging AC	Laying AC pipes	Cast-iron fittings	Role games	Wednesday	Preparing tap sites	Steel pipes	Tanks	
Thursday	Trench digging PVC	Visit to Phalombe and Sombani projects	Visit	Role games	Thursday	Preparing tap sites	Visit to Mulanje West and Namitambo projects	Visit	
Friday	Trench	Laying AC pipes	Cast-iron fittings		Friday	Tests	Tests	Tests	
Saturday	Removal of obstacles	Stores	Football		Saturday	Departure			

* AC = asbestos cement

second week each group practises the laying of asbestos cement pipes in their trench until the procedure is perfect.

The middle session, from 9.30 a.m. to 12 noon, is used for practical instruction in all aspects of handling and connecting asbestos cement, PVC, and steel pipes and fittings. Later in the second week the candidates are taught how to construct a standpipe apron, and a competition is held at the end of the course to discover the group with the best apron.

The afternoon session, from 1 p.m. to 3 p.m., is the "classroom" period, for lessons in project and community organization, stores, reading aerial photographs, writing work programmes, reports, etc. Study visits are also made to current and completed projects.

For each lesson or practical session, the candidates are given a lesson plan for their own record so that they do not need to make their own notes. On completion of the course these lesson plans constitute the Field Handbook (see Table 4.3) which they keep for the rest of their service.

At the end of the course candidates undergo practical tests which are marked by the instructors. These marks, and the overall performance of candidates, are discussed by all instructors, and candidates that are considered to have reached the standard required are selected. Sub-standard candidates are not selected, even if this means leaving vacancies unfilled. Successful candidates are appointed within 2 weeks of completing the course.

5.6 On-the-job Training

Training period

Any training course, however long and thorough, can never be more than initial training. It is a common fallacy that all that is necessary to produce the required manpower is to conduct enough training courses. In reality, a person who has been trained in this way probably only has about one-quarter of the training he needs. The poor performance of some programmes is due partially to the assumption that staff who have undergone a training course are fully competent, and can therefore be sent off to a project on their own.

It is essential therefore that field technicians undergo a further period of on-the-job training under close supervision before they are considered competent to work with less supervision. During this period they develop the technical skills learnt on the course, and the social skills *which can only be learnt by experience*. They should work alongside more experienced colleagues, and should be given specific attention and support by their Supervisors. Regular staff meetings play an important part in the training process.

In Malawi this training period lasted 2 years, after which, if their performance was satisfactory, the field technicians' appointments were confirmed.

Annual refresher/upgrading course

In addition to on-the-job training, field technicians should be given the opportunity to update their skills, discuss their problems, share their experiences with their colleagues and Supervisors, and learn new skills. An annual course should therefore be held in a slack work period. This course is also a social occasion that plays a valuable role in the maintenance of team spirit and motivation. It is particularly valuable for those more experienced field technicians working on smaller projects in relative isolation from their colleagues. Compared with those under close supervision, they may not be so aware of new procedures, new ideas, and new skills that are always developing on any programme.

The course also functions as an upgrading course. All staff should be required to take a test before moving up to a higher grade. The course will therefore have periods of formal instruction as well as periods of informal discussion.

5.7 Field Technicians' Career Structure

One of the most important factors affecting motivation and morale in any organization is a career structure with reasonable prospects of advancement. This is common knowledge, yet it is a fact that in some developing countries many government-employed staff have no promotion prospects at all, but are expected to stay at the same level for the whole of their career, with periodic salary increments. This is particularly true of relatively junior staff and field extension workers.

It is not easy for a programme or ministry to change government's long-established personnel regulations. However, by careful study of the rules it may be possible to identify a suitable category of staff into which field technicians can fit reasonably well.

In Malawi, the Project Assistants were originally employed on a temporary basis, being paid from project funds rather than from the government payroll. Unfortunately, because of government regulations, they could not be promoted to Supervisor unless they had at least 2 years of secondary education. Hardly any Project Assistants had this qualification, so however competent and experienced they became, the majority had no career prospects at all. This was not seen as a serious problem until the time came to promote some of the better ones to Supervisor level. If this could not be done, the whole organizational structure would become meaningless—experienced staff cannot supervise less experienced staff unless they are senior to them.

It was therefore necessary to find an alternative system. It was decided that Project Assistants should be reclassified as skilled workers, for which there existed a system of grades according to experience and skill, regardless of educational qualification. These employees could achieve a reasonable wage level compared with field workers on the permanent list. The Project Assistants were renamed Water Project Operators (WPO) with the following basic career structure:

(1) Selection after 2-week field training course.
(2) One to two years as ungraded WPO trainee.
(3) Two-week refresher/upgrading course and Grade III test.
(4) One to two years as WPO Grade III.
(5) Two-week refresher/upgrading course and Grade II test.
(6) Two to three years as WPO Grade II.
(7) Two-week refresher/upgrading course and Grade I test.

Field staff

(8) Rest of service as WPO Grade I (Supervisor).
(9) Grade I WPOs with appropriate educational qualifications (2 years secondary level) are eligible for selection for Foreman's course.
(10) Foreman's course and Test—successful candidates promoted to Foreman (equivalent to Technical Officer in the permanent list).

It is likely that classification of field technicians as skilled workers is preferable in most programmes, because this is usually the only way to avoid the requirement of having some secondary education. It has been mentioned above that primary-level field technicians are generally preferable in rural programmes because secondary-level people tend to have higher aspirations and often no longer identify themselves with the rural people.

However, the classification does not do full justice to the skills and status of field technicians who have greater responsibilities, require social as well as technical skills, and generally live and work in more difficult conditions than skilled workers in other ministries or departments. To cater for this, it is desirable to award some form of special 'field' or 'project' allowance on top of this basic salary.

5.8 Motivation of Staff

Promotion prospects and a decent wage are essential, but they are not the only factors that motivate people. Visitors to the programme in Malawi have been struck by the motivation and personal involvement shown by staff at all levels (WHO/World Bank, 1978. *Water Supply and Sewerage Sector Study*. Report to the Government of Malawi), yet until recently there was neither a particularly good career structure nor an adequate wage. It is difficult to analyse in detail how this degree of motivation has been achieved, but it is such an important factor in the success of any rural development programme that it is worth attempting to identify at least some of the factors involved.

The motivation process

The most important characteristic of motivation is that it is "infectious". Yet the degree to which an individual motivates others is dependent on the development process. If, for example, a motivated individual tries to launch a large-scale programme involving numerous field staff over a wide area it is unlikely that his own motivation will be sufficient to motivate them. If, on the other hand, he works with a small group of people in a Pilot Phase he will be closer to his staff and more likely to motivate a reasonable proportion of them. Those who are not motivated should leave the group, and will generally do so of their own accord. The remaining small group of motivated individuals can then carefully select suitable people to join them and, assuming a rigorous selection process, these would themselves become motivated and absorbed into the team. This process can be repeated until a formidable, highly motivated team is *gradually* built up.

This, in rather simplistic terms, is an interpretation of the motivation process that occurred in the Malawi programme. In reality the process takes time and patience, but the result can be striking. The 'take-off' point in Malawi came in 1972 when a nucleus of five motivated people very carefully selected 20 new individuals and succeeded in motivating them. Since then the process has been self-generating.

Factors affecting motivation

Some of the factors that create favourable conditions for motivation can be identified as follows:

(1) *Initial impressions.* Impressions and attitudes formed during initial selection, training, and early contacts with programme staff generally set the standard that a new staff member can readily adopt. If mediocre standards are initially acceptable it will be much more difficult to raise standards later.
(2) *Belonging to a team.* Team spirit can be fostered by regular meetings, open consultations between all levels of staff, sporting and social events, newsletters, plotting team progress, friendly competition, etc.
(3) *Remuneration and promotion prospects.* A necessary but not sufficient condition for motivation.
(4) *The loyalty and support shown by Supervisors.* Supervisors who are concerned for the welfare of their subordinates, who actively promote their cause in justifiably seeking better conditions, and who regularly visit them in their place of work will in turn elicit their loyalty. "Absentee Supervisors" rarely motivate anyone. This applies not only to immediate Supervisors but also to senior ministry officials.

Figure 5.1 A happy group of Project Assistants after successfully pressure-testing their section of 200 mm pipeline

(Photo: the author)

(5) *A specific objective.* It is generally easier to be motivated in a job that has a specific objective, such as to construct a water supply system, than in a routine job, such as accounting or clerical work. Constructing a water system is a dynamic process in which different techniques and procedures are applied as the work progresses. This contrasts with the repetitive nature of many other jobs.
(6) *Responsibility.* A certain degree of responsibility is an essential condition for motivation. Provided responsibility is given within a framework of sufficient support and supervision, field staff whose talents may otherwise have passed unnoticed can succeed even in the most difficult tasks. For example, in Malawi, a Project Assistant excavated a tank site 15 m in diameter and 2 m deep out of fractured rock, by skilfully managing and motivating a self-help labour programme lasting several months; the Project Manager (the author) originally thought it was impossible without explosives and excavation equipment.
(7) *Social pressure.* In a programme which involves and responds to the needs of the community, field staff who live and work with the community naturally come under social pressure to carry out their duties conscientiously.

5.9 Supervisors

Initially in a new programme it is necessary to recruit Supervisors from another rural development programme. It is essential that they have had good experience of rural conditions, particularly with working in rural communities. Their technical qualifications are less important than their personal qualities. They should be competent and confident enough to supervise staff and should therefore be very carefully selected by the senior programme staff and not, as often happens in a civil service, arbitrarily transferred from another department without consideration as to suitability. Ideally such Supervisors should initially be employed on a temporary basis, so that they can revert to their original department if they turn out not to be suitable.

As the programme develops, Supervisors should be drawn from the ranks of the field technicians. Not only is this an essential feature of their career structure, but experienced field technicians make the best Supervisors. They should have a minimum of 5 years experience, though accelerated promotion may be necessary in the early period of the programme. Their selection should be on the basis of an assessment of their field performance, coupled with some competitive tests or interviews. If a career structure similar to that described in Section 5.7 is used, Grade II field technicians with 2 or 3 years experience in this grade could be promoted to Grade I (Supervisor) after a 2-week upgrading course and grade test. Generally, Supervisors do not need any special training other than this upgrading course.

The ratio of Supervisors to field technicians will depend largely on the nature of the programme. On average, each field technician should meet his Supervisor at least 1 working day in every 5. In a large-scale project with easy communications, this may mean he is visited once every 10 days, and attends a staff meeting once every 10 days, giving an average contact frequency of once every 5 days. In a programme involving a number of smaller projects scattered in difficult terrain, regular staff meetings are impractical and the Supervisor's visit would need to last more than 1 day at a time. In this case the Supervisor may visit each technician 1 week in every 5. In all cases the frequency and intensity of supervision will vary according to the tasks in hand and the experience of the technician. More supervision is required at the beginning of a project, when the self-help labour programme is being tried for the first time and when intake structures are being constructed, than later during routine trench digging or pipelaying. Similarly, more supervision is required at the end of the project to make sure all the details are finished off properly.

In general, therefore, the ratio of Supervisor to field technician will vary between 1 in 10 for an easily supervised programme, and 1 in 5 for a more difficult one.

In addition to field supervision, Supervisors often have other responsibilities. Experienced Supervisors are required for feasibility surveys, particularly to assess the *social* feasibility of proposed projects; they are required as instructors for training courses; they may be given special responsibility for stores, to ensure the storekeeper is keeping proper records and that stores are well looked after. Much of the Supervisor's time must be spent in "trouble-shooting"—concentrating where the local community is divided, helping out a trainee technician who is in trouble with the community, or visiting a particularly difficult section of trench that is being dug.

5.10 Engineers

Role

The principal role of the engineer is to be the leader of the team. He may be supervising a group of smaller projects within a local area, or he may be the Project Manager of a major project. A sample Job Description is shown in Appendix 2.

He carries out the detailed surveys and designs, submits material requirements, plans the implementation schedule, motivates and coordinates the self-help labour programme, supervises field staff and their training, supervises the work of skilled contractors, inspects the projects on completion, and liaises with other government officials. His duties are mainly technical and managerial but, as with all field staff, it is essential that he is sensitive to community and social issues and that he develops some skills in committee work.

Recruitment

As with the lower levels of staff, very careful selection is required. Engineers who have been working in traditional technology-oriented jobs may find difficulty in adjusting to the flexible, community-oriented and low-cost technology approach. On the other hand, engineers fresh from college are generally attracted to the sophisticated, high-cost technology that they studied and the best of them will tend to get jobs in that field. Rural development work is often the last choice of most engineers seeking a job. The problem is aggravated by the fact that in most developing countries there is a shortage of trained engineers.

One way of solving the problem is initially to recruit diploma engineers rather than degree engineers. Their training is generally less sophisticated and less academic, and they are often more ready to learn and adapt to engineering in rural areas. This is not to say that engineering in rural development work is inferior to sophisticated technology work. In fact simple, low-cost *community-oriented* technologies are often more of an engineering challenge. The engineer is left much more to his own ingenuity and his engineering appreciation of the situation. Almost certainly, a young engineer in rural development work will get much more responsibility and first-hand experience than will his counterpart in the high technology sector.

The most effective way of attracting recruits is to contact the college from which the engineers are about to graduate and arrange a study visit to see the work on a particular project. This gives potential candidates an idea of the work, and can also give them a preliminary impression of the high morale and team spirit of the programme staff. This is often an attractive feature and at least a few of them will become interested. These should be invited to fill out application forms to apply for the vacant posts. Applicants should then be interviewed according to normal government procedure and successful candidates should initially be given a probationary appointment which will be confirmed after satisfactory performance.

Training

Newly appointed engineers should be sent for initial training, preferably to work on a major project as an assistant to the Project Manager, or to work alongside the engineer in charge of a number of smaller projects. They should follow a training programme, not just work haphazardly from day to day, so that they become familiar with all aspects of project work. In particular they should spend a period working with a field technician and with a Supervisor. They should also be given certain technical tasks such as survey and site work. This field training period should last about 3 months, or less if the engineer has appropriate previous experience.

After the field training they should spend a further 3 months at the programme headquarters to carry out design work and learn the administrative, logistic, and accounting systems. Again, they should have a specific training programme and be closely supervised by the head of the programme or another experienced engineer.

After this period of training they should be appointed either as assistant engineer on a large project, or as the engineer in charge of a number of smaller projects. It is essential that inexperienced engineers are given adequate supervision and support, and that they are able to rely on experienced Supervisors and field technicians.

Further training

A rural water supply programme must have a certain number of fully qualified professional engineers if it is to develop along sound professional lines and be accorded equal status with the other professional ministries and departments. The best way to achieve professional standards while retaining the community-oriented engineering approach is to send some diploma engineers, after a suitable period of field experience, to a professional degree course. Professional engineers brought in from other programmes who have not had the benefit of the unique experience in the particular programme will usually be unable to understand and adapt to the different approach required. The possibility of a degree course also acts as an incentive to the diploma engineers both to join the programme in the first place, and to perform well. Once suitably qualified, the professional engineers can return to the programme, often in relatively senior positions. Again it should be stressed that academic qualifications alone do not guarantee the competence of an engineer. Non-technical administrators and senior civil servants tend to assume that once an engineer or technician has completed an appropriate course of training he is automatically competent to carry out his duties. In fact, his *previous experience* will have a much greater bearing on his competence.

5.11 Conclusion

The success of any rural development programme depends heavily on the quality and training of field staff. One of the problems experienced in many countries has been finding the appropriate employment structure within a civil service whose personnel policies are more geared to urban-oriented institutional requirements than to the needs of rural development. In addition, government policies often encourage the attainment of educational qualifications by raising educational standards for Civil Service posts. Paradoxically, this militates against rural development, for the education system is such that higher education is usually synonymous with greater urban orientation. Meanwhile, there is a great reservoir of talent and ability among people in the rural areas with relatively lower educational standards who have no wish to join the urban unemployed and who welcome the chance of rural employment. Such people, with suitable training and motivation, can become effective agents of rural development.

CHAPTER 6
Technical aspects

6.1 Introduction

Water supply is a field in which the engineering profession can make the most immediate contribution to the quality of life of rural societies. The experience of the rural water programme in Malawi shows how engineering and social disciplines can be combined to construct socially acceptable and beneficial water supplies that are also technically robust and efficient. Most of this book is concerned with the non-technical aspects of a rural water supply programme, as these are definitely the most crucial and the most difficult to develop. The technology that is used, however, is also of fundamental importance to the success of the programme—the use of inappropriate technology will seriously affect the programme, no matter how well-organized and managed it may be.

This is not a technical textbook, so no attempt is made here to cover all the different technologies that are used in different countries, or even within countries. Instead, a fairly detailed description of the technology used in the rural piped water programme in Malawi is given, as a practical example of the types of technical features that are important. This account may also be useful as one of the very few first-hand technical descriptions of a rural water programme.

6.2 Design

Design criteria

The design of the first project (Chingale 1968) was based on three simple criteria:

(1) Design consumption of 27 litres (originally 6 gallons) per head per day.
(2) Design flow of 0.075 litres per second (originally 1 gallon per minute) per tap, assuming all taps open at once.
(3) Allocation of 1 tap per 300 population (i.e. up to 300, one tap; 301–600 two taps; 601–900, three taps, etc.)

The first two criteria are still in use today, and are discussed below. The third, however, proved to be socially unfair and technically inconsistent. The social problem is illustrated by the fact that two villages of the same apparent size with populations just below and just above 300 would theoretically have one and two taps respectively. The technical problem became apparent on the first major project, for which the initial allocation on this basis indicated a figure of 300 taps to serve 75,000 people, an average of 250 persons per tap. This average was inconsistent with the first two criteria, which indicated that even if the tap was delivering water at the design rate for an unrealistic 24 hours, it could only serve 240 people.

During installation, the project engineer received numerous requests from villages for extra taps, many of which, on examination, proved to be reasonable. The extra taps were installed, leading to a total of 460 taps with an average of one per 163 persons.

The lesson for subsequent projects was that much greater attention to detail was necessary at the design stage. Most of the extra taps required could have been foreseen in the light of subsequent experience. The third criterion was therefore modified to allow the allocation of one tap per 160 persons, and a fourth criterion was introduced to the effect that the design is based on the assumption of 16 hours service and 8 hours storage within any 24-hour period. The design criteria currently in use are given in Appendix 3.

Integration of design with construction

It is a significant feature of the Malawi programme that engineers design their own projects. This contrasts with the more common practice of separating the functions of design and construction. Although this separation may be unavoidable in complex, high-technology projects, it is counterproductive in simple technology, field-oriented schemes. The advantages of integrating design and construction lie in the automatic feedback of field experience to show up inadequacies in the design, in the extra flexibility and speed with which design can be adjusted to requirements, and in the motivating effect on engineers who identify themselves personally with both the design and the construction processes.

In Malawi the project engineer as the designer is in very close contact with the field staff, gaining the benefit of feedback from their experience as well as from his own observations. Thus modifications to design details and construction procedures are easily carried out, and the effectiveness of design procedures is under continuous observation.

Design procedure

To ensure the standardization of designs and to enable less experienced staff to design relatively large water supply systems in Malawi, the design procedure has been laid down in considerable detail in a special handbook. This procedure is described in Appendix 3.

Design consumption

The design consumption of 27 litres per capita per day appears to be a reasonable figure both in the light of present consumption and the experience from elsewhere in Africa. It is clear from the operation of completed supplies that present consumption is much less than this, although it has not been accurately measured. An estimate carried out on a branch line of one project in 1976, based on the water drawn from a branch storage tank, indicated the consumption was about 16 litres per capita for a particular 24-hour period. Feachem *et al.* (1978; *Water, Health and Development* (Tri-Med Books, London), p. 107) report a similar average consumption figure for rural communities in Lesotho.

Design life

The design procedure described in Appendix 3 does not refer explicitly to the design life of the system. Instead, the system capacity is based on a design population, which is an estimate of the maximum population that the area can support with traditional agricultural methods. This concept assumes that as the pressure on the land increases, migration will occur from the over-populated to the under-populated areas of the country.

This assumption is partly true, as there is evidence in Malawi, for example, that within Mulanje District people have moved from the densely populated Mulanje West to the less densely populated Phalombe area following the installation of a water supply there. However, migration is unlikely to account for all the population growth in a densely populated area, and population densities will continue to rise. The pressure on land may encourage the adoption of more modern methods of agriculture, including the introduction of higher-yielding crop varieties, so that the land will be able to support a higher population.

An alternative approach is to specify a desired design life, and estimate the population growth and per capita consumption growth to arrive at the system's design capacity. This method appears to be more logical, but its effectiveness depends heavily on the accuracy of the estimates. The national figure for population growth rate could be grossly inaccurate when applied to a project area, and the growth rate of per capita consumption is dependent on so many factors, including the design figure itself, that any estimate is little more than guesswork.

It would seem, therefore, that the concept of design life for rural water supplies in developing countries is fraught with so many imponderable variables and implications as to become almost meaningless. A more practical approach is to decide on a consumption figure that allows a reasonable margin for growth without requiring an excessive capacity beyond the ability of the programme and the community to install; and to decide on a reasonable design population, based partly on the natural population growth rate and partly on the capacity of the land to support the population. This method is certainly not perfect, relying as it does on intelligent estimates and guesswork, but it is probably as effective as any other, and would appear to be reasonable for the situation in Malawi.

Average flow

From Appendix 3 it can be seen that the design procedure makes no allowance for peak factors. The design assumes that each tap is continuously in service for 16 hours at a constant flow rate. As a rule of thumb, this procedure appears to have satisfactory results, although it is clearly unrealistic. This might be explained by the fact that the system is designed for the theoretical condition that all taps on the supply are in use at the same time. It would appear that this condition does not occur, because even at the peak demand periods (6 a.m. and 6 p.m.), the flow rate at any particular tap is usually more than twice the design flow rate.

Despite the fact that the system appears to work satisfactorily without any peak factors, there is a need for more research into the effectiveness of the design, to indicate what improvements should be made. At present, very little is known about actual demand or its daily and seasonal variations. Although it is unusual to see more than two or three women waiting at a tap, it is possible that as demand approaches the design consumption figure queues will form during the peak periods.

Communicating the design

Field technicians will have had relatively modest formal education and it is therefore important that the design of the water supply system is presented in a simple and easily understood format.

On the first major project in Malawi, the engineer originally introduced an elaborate system of symbols, colour codes, fittings charts, diagrams, maps and aerial photographs. This system confused the Project Assistants, because essential information was obscured in the wealth of detail and because the same information was duplicated in different forms. However, the experience did show that they had little difficulty in learning to interpret *aerial photographs.*

The present system uses aerial photographs only, suitably marked with coloured chinagraph pencil to convey all essential information. The photographs show the pipe alignment; pipe sizes; the positions of valves, standpipes, and tanks; and the positions, names, and populations of villages. Three sets of photographs are marked up at the design stage, one for issue to Project Assistants, one for display in the project headquarters, and one for the use of the engineer.

Ordnance Survey maps are also marked up with similar information but do not normally need to be referred to by Project Assistants. They are mainly for the use of the engineers and programme headquarters. Examples of the map and aerial photograph can be seen in Appendices 4 and 5.

6.3 Construction Techniques

It is most likely that a new programme will start with technical designs which are relatively untried in that particular environment. Some of the technical aspects will be tested in the Pilot Phase, but not all. There will be several modifications required during the course of the programme as experience is gained and weaknesses are shown up.

Some design modifications will have little significance at the field level, such as a decision to increase storage or spare capacity for future development. Others will have direct impact on the field staff, such as details of standpipe aprons or soakaway pits.

In the Pilot Phase in Malawi it was vital to establish the fundamental principles that water could flow long distances in pipes and that the Government would provide the pipes if the people would provide the labour. These principles were so important that technical details were of a lower priority. Once these principles were established and confidence gained, the standard of installation became the priority. This may not be a valid generalization for any programme but it does focus on some of the fundamental goals of the Pilot Phase, and how they may differ from the goals of the subsequent programme.

Technical aspects

In the pilot projects, therefore, the main concern was to motivate the people, to install the pipeline, and deliver water at the taps. The design concentrated only on the selection of pipe sizes to ensure adequate and even distribution of water. Standpipe aprons consisted merely of a rough collection of stones, and the tap was usually connected before the drain and soakaway had been dug.

For later projects the recruitment of Project Assistants made it possible to improve the technical standards of installation. The village tap was not connected until a proper apron, drain, and soakaway had been dug and all other aspects of the supply completed to the required standard. Early experience with storm damage led to particular emphasis on the protection of open trenches and completed pipelines from erosion.

6.4 Technical Description of a Major Project

The technical simplicity of a relatively large-scale project can best be conveyed by a detailed description of the system as it operates. As an example, the Phalombe Water Project is described here. This supply was completed in 1977, serving 90,000 people in 135 villages through 578 standpipes. It covers an area of about 625 km^2 and involves 56 km of asbestos cement and 435 km of PVC piping.

Intake

Water is drawn from a natural pool in a river flowing off Mulanje Mountain. The pool is situated about 500 m above the forest boundary. Five 100 mm galvanized steel pipes are set in a small concrete weir across the original outlet of the pool. (Larger-diameter steel pipes are difficult to manhandle and less readily obtainable.) Each intake pipe is closed at its end, and 600 holes of 9 mm diameter are drilled over a length of about 1.2 m. These holes have a combined surface area several times the cross-sectional area of the pipe, thus reducing suction. The number of holes means that the supply is unaffected by a single flood, but in the rainy season successive floods will gradually cause the holes to become blocked with debris. A watchman is therefore employed to keep the holes clear (see Section 7.4).

Screening tank

The purpose of this tank is to remove the pebbles, sand, twigs, leaves, and grass which have passed through the intake pipes.

The tank is situated about 100 m from the intake. It has an overall size of 2.8 m long × 1.9 m wide × 1.2 m high, and is covered with removable slabs. There are three chambers: the first is an upward flow chamber to settle out heavier matter such as stones and sand; the second is a horizontal flow chamber containing two vertical 6 mm wire-mesh screens on removable frames; and the third is the outlet chamber, separated by a 0.9 m high baffle wall. The screens are cleaned daily in the rainy season as part of the watchman's maintenance routine, although they usually survive two or more floods before becoming blocked. (More recent experience in Malawi suggests that the screening tank is unnecessary as its function can be carried out by the sedimentation tank.)

Figure 6.1 An intake using a natural pool in the river—these four 4-inch pipes serve 90,000 people. The intake is built within a forest boundary which is patrolled by forest rangers

(Photo: the author)

Figure 6.2 An intake must be strongly built

(Photo: the author)

Technical aspects

Sedimentation tank

This is designed to give a mean retention period of about 2 hours to allow fine matter to settle. Jar tests showed that clarification is complete in about 1 hour. The tank is also the main header for the whole supply system and is situated at an elevation of about 80 m above the lowest point in the first pressure stage. For topographical reasons it is about 1.6 km from the screening tank, connected by an asbestos cement main in sections of 150, 125 and 100 mm pipe, calculated to give exactly the right flow to the tank. The tank is a circular, reinforced concrete structure with a fixed roof. It has a diameter of 12.8 m, water level height of 1.8 m and volume of 227.5 m^3 (50,000 gallons).

The inflow first enters at the bottom of the tank into a baffled inlet chamber, and then rises up into the main body of the tank. The outlet is also at the bottom of the tank, separated by a 0.9 m baffle wall. This allows about 1 hour's reserve storage with no inflow (or several hours with reduced inflow) and avoids air being sucked into the main line should the inflow be reduced. Despite the provision of air valves in the main, air entering the pipe can affect the flow considerably, and the effect may be felt for some time later at remote parts of the supply. The inlet supply can sometimes vary, as it does when the intake or screening tank is being cleaned, or if the watchman neglects his duty after a flood. So the use of the sedimentation tank to provide automatic emergency storage reduces the supply's vulnerability to inflow variations. The temporary reduction in retention time is less harmful than the ingress of air into the system. The problem could also be solved by building a second tank as a balancing storage tank, but the expense of this is not justified. Under normal flow conditions the tank is kept full by the fact that the inflow exceeds the outflow, the excess passing through a bellmouth overflow set at the desired water height.

The tank is fitted with a by-pass from inlet chamber to outlet chamber so that the supply can be maintained without interruption while the tank is emptied for cleaning. This avoids having to have duplicate facilities, but it means that the cleaning of the tank must not be done while the river is turbid, as there is clearly no sedimentation while the by-pass is being used.

On Phalombe, the sedimentation tank is fitted with an outlet sluice valve. However, this is not recommended, as the rapid operation of this valve causes surge pressures which can lead to multiple pipe bursts in the main. Since operation by unskilled personnel is a strong possibility over the lifetime of the supply, the sluice valve is likely to cause more problems than it avoids. In the event of a burst the main supply can be cut off by closing the *inlet* sluice valve and opening scour pipes to empty the tank. In reality the tank would already be empty by the time the sedimentation tank was reached after a burst.

A circular sedimentation tank is not the most efficient shape from a technical standpoint because of the degree of short-circuiting. Ideal plug-flow conditions are best approached by a long rectangular tank. However, from a management viewpoint it is easier for the construction team if all tanks, both storage and sedimentation, are to the same basic design. A large rectangular tank is inherently more difficult to construct to a good standard than a circular tank, especially for semi-skilled rural artisans, and as only one sedimentation tank is built per project compared with several storage tanks, their unfamiliarity with the rectangular tank would lead to poor-quality construction. This is an example of how normal technical practice has to be adapted to meet the local situation.

Main pipeline

The supply is divided into six areas, each one being supplied from a main storage tank (227.5 m^3). The main line supplies these storage tanks, although some minor reticulation is installed direct from the main to serve the intermediate areas.

The main begins from the tank in 225 mm Class 12 (maximum working pressure 6 kg/cm^2) asbestos cement pipe and reduces in size progressively through 200, 150, 125, and 100 mm as it proceeds through the areas. Class 18 is used for those sections with higher maximum pressure. The mixing of classes caused problems with fittings as the exterior diameters varied between classes. Ideally only one class of pipe should be used and, although this leads to higher costs, the overall cost-effectiveness still compares very favourably with other systems.

The pipes are buried in a trench 1.2 m deep. The completed lines are pressure-tested using a hand pump to 75 per cent of the manufacturer's test pressure, or 150 per cent of the maximum static pressure, whichever is the smaller.

Double or single air valves are fitted at all high points in the profile or sudden changes in gradient, in addition to a double air valve situated every 3 km.

Flushing valves are fitted at all low points in the profile to clean out sediment and to flush the line after the repair of a burst. All valves are protected by standard concrete culvert rings placed on end and covered. These are superior to brick valve-boxes and easier to install.

Where the main crosses a river or stream, it is usually buried well under the bed and protected with a checkdam to stop erosion. This is a very low weir built just downstream so that silt is deposited over the pipe. Where it is impossible or impractical to dig under the bed, steel pipes are supported above the maximum flood level on pillars. On two major projects, such a pipe crossing has been constructed with special railings and boards to serve as a footbridge. These river and stream crossings are potentially vulnerable spots and receive special attention in the maintenance inspection routine (see Chapter 7).

Storage tanks

The area storage tanks are at the end of the first pressure stage, usually about 60–90 m below the sedimentation/header tank. They act as balancing tanks so that the main line is utilized at maximum design flow continuously for 24 hours a day, the tank drawing down in the daytime and filling up again during the night. Each is fitted with an equilibrium float valve so that the inlet closes when the tank is full. This avoids the wastage of water that would occur if one tank overflows while others are still filling elsewhere.

Branch lines

The area storage tanks supply a second-stage trunk main which itself supplies a number of branch lines. Most branches have their own branch storage tank where

Technical aspects

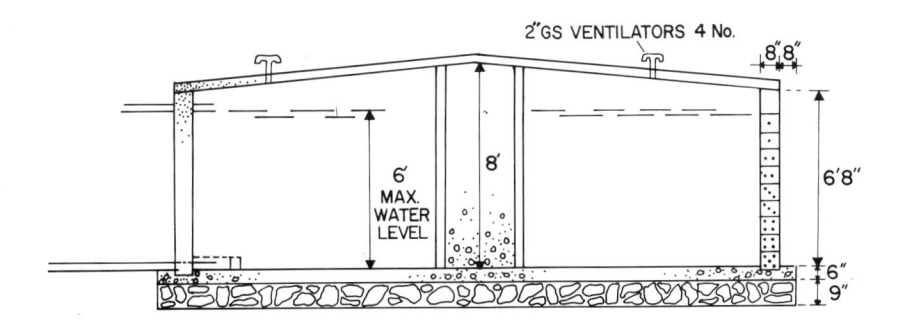

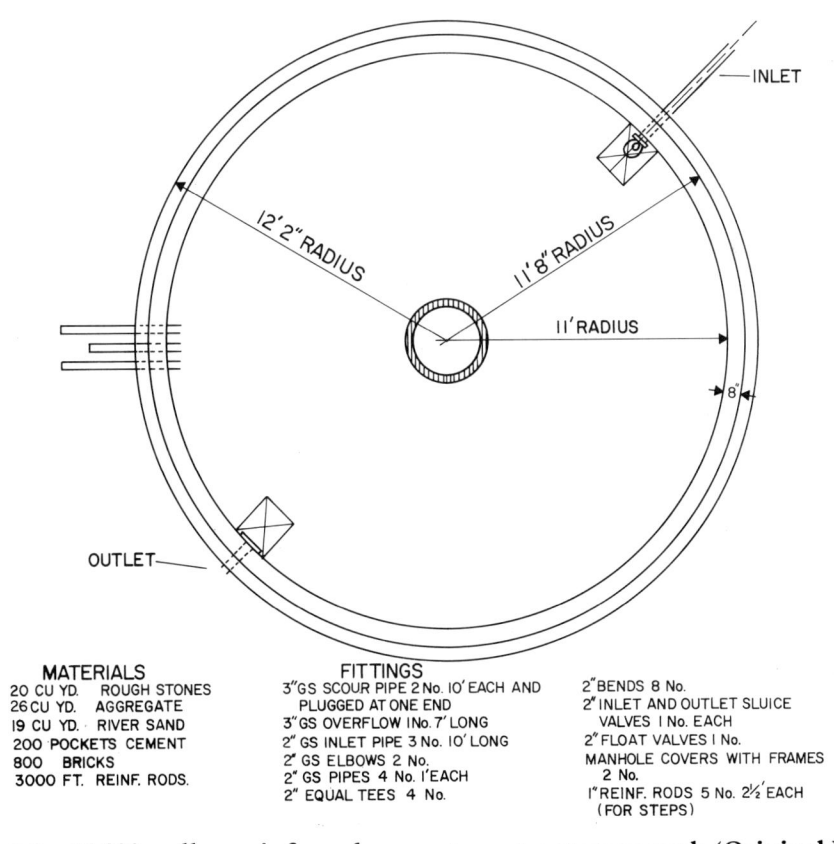

Figure 6.3 15,000-gallon reinforced concrete water storage tank (Original Imperial units shown)

Figure 6.4 Village tap apron (Original Imperial units shown)

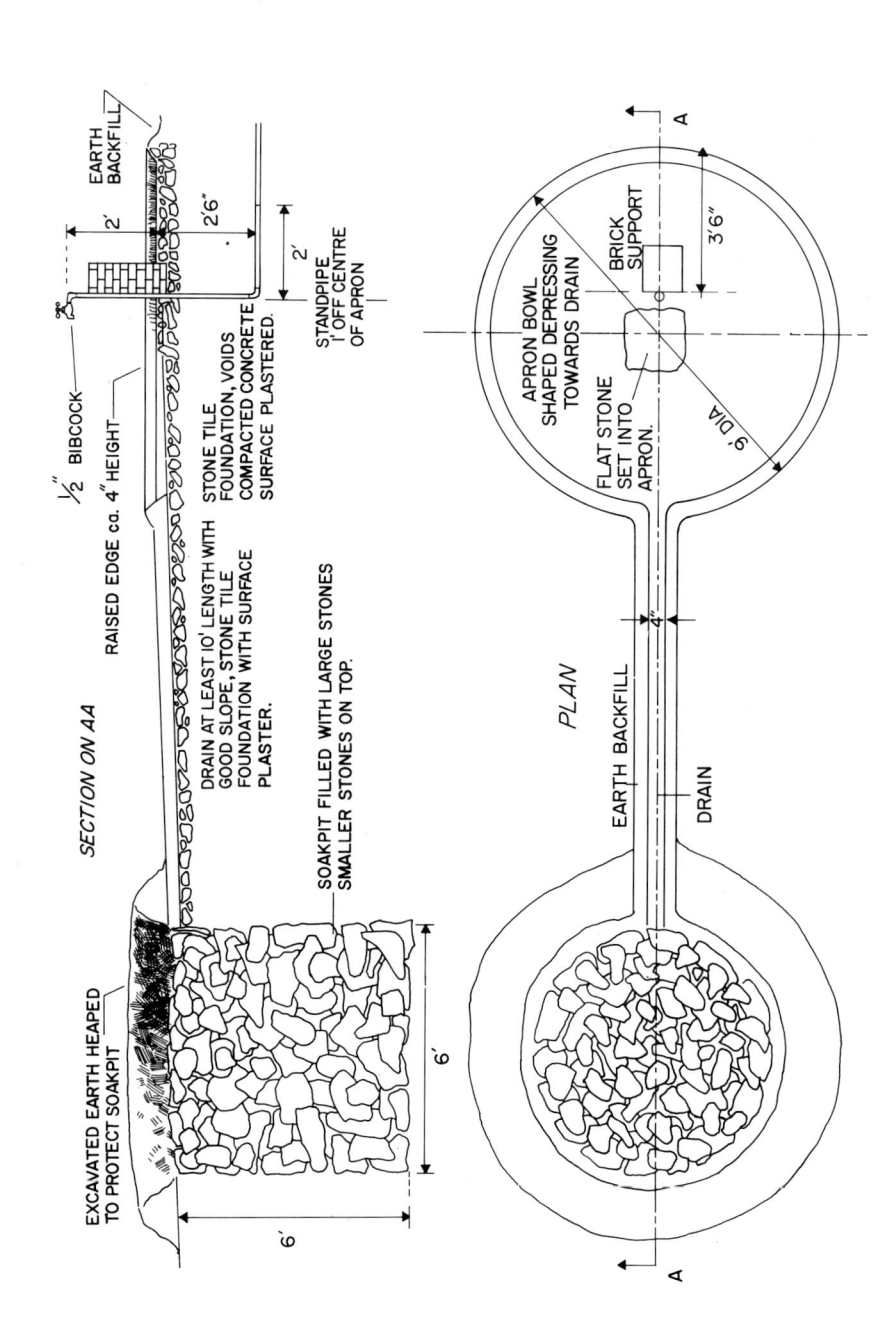

the topography permits. Pipe sizes get progressively smaller according to the population served, so that equal distribution is effected by the design. Most of the second-stage trunk main is in asbestos cement, being 100 mm or larger. Most branch lines are 100 mm and less and these are laid in rigid PVC at a depth of 0.75 m (2.5 ft). Gate valves are situated at every junction or size change involving 40 mm and above, and at least every 1.6 km for all pipe sizes. The valves are protected by a 0.75 m length of 100 mm asbestos cement pipe placed vertically over the valve, with a specially made cover that slides over the top of the pipe. These lengths are readily available from offcuts of damaged pipes. The diameter of the pipe is large enough for an adult arm, but the length is too long for children to reach the handle of the valve. Nevertheless, the valves seem to be a temptation for a few inquisitive fiddlers, and so the valve handles are now removed and given for safe keeping to the nearest householder or to the village headman or committee member. It only takes one such fiddler to disrupt the supply and, unless the valve is closed completely, it is quite a tedious problem for the Maintenance Assistant to check that all gate valves in the affected area are fully open.

All pipelines are marked by a ridge and planted with drought-resistant *Paspalum* grass. Particular attention is paid to streams and gully crossings which are major erosion hazards. The ridge is discontinued over the gully to allow surface water to cross the pipeline and a small checkdam is built just downstream so that silt is deposited over the line.

Village taps

For most people, the village tap is the most important part of the whole project. If it is not to become an unpleasant muddy mess it must have a proper apron, drain, and soakaway, and the standpipe and apron must be particularly robust if they are to survive the heavy use to which they are subjected. The Malawi programme has developed a standard Village Tap Site which has to be *completed before the tap is issued*. The standpipe consists of 15 mm (½ inch) galvanized steel pipe with a 15 mm brass crutch-headed bibcock (tap). The PVC supply line is first connected to a 0.6 m horizontal length of 15 mm steel pipe which is then connected to a 1.4 m vertical length as the standpipe. The tap is thus 0.6 m above ground level. The standpipe is supported for its full length by a 75 mm diameter de-barked hardwood post cut locally, to which it is tied with galvanized wire.

The standpipe is slightly offset from the centre of a 2.7 m diameter saucer-shaped apron which has a depression towards the drain. The apron is constructed on a foundation of stone or broken brick which is compacted with concrete to fill the voids. The surface is rendered with a cement–sand mix. A large flat stone is concreted into position under the tap to act as a stand for the bucket and to take the impact of water which would otherwise erode the concrete.

The apron is connected to a soakaway pit by a drain at least 3 m long. The soakaway pit is circular, 1.8 m diameter and 1.8 m deep, and filled with large stones. Earth is heaped around the pit and edges of the apron and drain to stop surface water causing silting problems. Each tap site is issued with two 50 kg pockets of cement, which is sufficient to complete the apron and drain.

90 Village water supply in the decade

Figure 6.5 A completed tapstand—note supporting post, anti-erosion stone under bucket, good-size apron, drain, and soakaway pit protected from surface water by earth ridge. This tap is about 20 km from the intake

(Photo: the author)

Figure 6.6 A similar tapstand from a different angle

(Photo: H. Van Schaik)

Technical aspects

Maintenance

Maintenance work on Phalombe is the responsibility of two Maintenance Assistants and one watchman employed by the programme. The Maintenance Assistants were formerly Project Assistants on the project staff. The watchman, who is responsible for the maintenance and surveillance of the headworks (intake, screening and sedimentation tanks), is a local villager. A description of maintenance routines is given in Chapter 7.

6.5 Asbestos Cement Pipes

Availability and cost

Asbestos cement pipes are usually cheaper to manufacture than plastic or steel pipes, except in smaller diameters. If a local manufacturer exists, this will probably be the most economical type of pipe to use for sizes 100 mm and larger. If there is no nearby source, the cost of transportation from a distant supplier can make absestos cement pipes prohibitively expensive. Malawi was fortunate to have a relatively close source in Mozambique situated on a direct rail link, and asbestos cement was usually used for all pipes of 100 mm diameter and larger. The use of asbestos cement pipes is, however, limited to programmes with reasonably easy internal transport facilities. They are bulky and heavy, and normally require a lorry to deliver them to site.

Handling and laying

Undoubtedly the biggest drawback in the use of asbestos cement pipes is the danger of hair-line cracks being formed during careless transport and handling. To reduce this risk, self-help labour is always used to handle the pipes from the railway trucks onto project transport, even though this means transporting the labour long distances. (Paid labour leads to a much higher incidence of careless handling.)

The only serious problems with asbestos cement pipes were experienced on one project in 1974 when an 8 km section of 225 mm Class 12 (6 kg/cm^2 working pressure) pipe yielded about 30 bursts under test. The remaining 48 km of pipe were relatively trouble-free. It is likely that the bursts were due to hair-line cracks sustained during transit some 40 km from the rail-head along "corrugated" earth roads and finally along the self-help temporary road by the trench. The latter section passed through gardens whose former ridges and furrows, although apparently levelled out by the self-help labour, gave the laden lorry a rough ride. This was too much for some of the Class 12 pipes in this relatively large size. No problems were experienced with the smaller pipes which were mostly Class 18. Class 18 (9 kg/cm^2) is therefore considered the lightest class suitable for transporting on rough roads in rural areas.

Although great care needs to be taken to level and prepare the bed of the trench, experience has shown that the high standards of laying required can be achieved by well-supervised teams, with relatively little training, supported by self-help labour. The actual joints themselves are the least difficult part of the laying process. In

Malawi, asbestos cement collars with rubber rings were used for jointing the pipes. Cast iron fittings were connected using metal couplings.

Pipeline movement

Experience in Malawi has shown that asbestos cement pipes can be used even in unstable soils and low-lying swamps. The heavy dark clay soils become waterlogged in the wet season and are prone to some vertical movement; in the dry season they crack severely. It seems there is sufficient flexibility at the pipe joints to cope with some vertical or lateral movement and the longitudinal stresses and strains caused by the soil cracking are either within the compressive and tensile strength of the pipes, or are balanced by the slight movement of the pipes relative to their joints.

Longitudinal movement also occurs, especially when the line is first put under pressure. Some movement may occur due to thermal expansion and contraction, though this is minimal under normal, constant water temperature conditions. However, if successive pipes are allowed to butt together inside the joints, a compressive load may be set up that can cause a crack leading to a burst. To avoid this, care must be taken during the laying to leave a gap of about 12 mm between each pipe-end. On larger sizes a central rubber "cushioning" ring is inserted in the collar so that the pipes can be pushed "home" without damaging the ends and compressive forces are then absorbed by the rubber itself.

6.6 PVC Pipes

Availability and cost

The relative cheapness and remarkable qualities of rigid PVC pipe have made possible large-scale reticulation in rural areas which was previously prohibitively expensive and technically impractical. One of the advantages of PVC is that it can relatively easily be manufactured in a developing country, compared, say, with asbestos cement or steel pipes. Even if they have to be imported, the freight costs can be kept relatively low by nesting the pipes.

The programme in Malawi used both imported and locally manufactured pipe in sizes 90 mm and below. In some cases larger PVC pipes have been used when asbestos cement pipes were not available or were inappropriate.

Handling and laying

One of the most significant properties of PVC is its low density. Huge quantities of PVC pipe can be transported in nested form, which permits relatively high-density packing and reduces transport costs. Pipes can be distributed to projects by road transport in compact efficient loads. For example, a 7-ton lorry with pipes stacked 1.8 m high can carry 12 km of 40 mm, and 12 km of 25 mm, Class 10 PVC pipe with a total weight of 6 tons. Within the project area pipes can be efficiently distributed by Land Rovers equipped with a special overhead frame, and then carried by head-load to the trenches. By comparison, the physical task of distributing, say, 400 km of steel pipe for a major project would be enormous.

Technical aspects

While PVC pipes need to be handled with care, they are far more robust than asbestos cement pipes, and it is rare for damage to be caused by careless handling. The jointing procedure using cleaning fluid and solvent cement is readily learned by rural people, who are familiar with the need for cleanliness and correct application for the repair of bicycle punctures. The alternative procedure using rubber rings needs closer supervision as the ring can easily be inserted the wrong way round. Although the process can be performed by village labour, pipe-laying should always be supervised by a field technician.

Sizes and classes

In Malawi, Class 10 (maximum working pressure 10 kg/cm^2) piping is most commonly used, although Class 6 (6 kg/cm^2) is occasionally used to reduce cost in the larger sizes (110 mm and above). To ensure equal distribution of water all pipe sizes are employed, namely, 12, 16, 20, 25, 32, 40, 50, 63, 75, 90, and 110 mm outside diameter; 20 mm is the normal minimum size, but 16 and 12 mm are used to reduce flow to a tap near a high pressure line. Above 110 mm, asbestos cement is usually preferred but PVC has been used in sizes 125, 140, and 160 mm in small quantities.

Effect of sunlight

PVC pipes are susceptible to ultraviolet rays which discolour them and make them more brittle. They then become more prone to damage during handling and are liable to burst at pressures below the designed maximum. In addition, the lower rows of a stack of pipes exposed to bright sunlight are liable to distort under the higher surface temperature and the weight of pipes above them.

The early projects in Malawi used imported pipes that were dark grey in colour. The manufacturers were then asked to supply white pipes in the hope that this would reduce their vulnerability to sunlight. This has dramatically reduced the surface temperature of pipes left in the sun, and has also reduced the discolouration effect but it has not been possible to determine whether the embrittling effect has also been reduced.

It should be stressed that PVC pipes should always be stored in the shade. On a major project a special shade is built, while on a minor project the pipes may be stored in a naturally shady place. However, when pipes are eventually distributed to the villages, they are inevitably exposed to sunlight for a while.

Longitudinal movement

PVC pipes laid during the heat of the day are naturally in a slightly elongated state. The water is turned on after backfilling the trench and the cooling causes them to contract in length. This effect caused some problems on one project where bursts in 110 mm and 90 mm occurred at some solvent cement joints when the spigot pulled out of the socket. Smaller sizes were unaffected. This was attributed to the fact that the larger pipes were more rigid, and therefore laid straighter, whereas the smaller pipes were "snaked" into the trench. The smaller pipes were able to cope with the tension by "straightening" themselves a little, whereas the larger ones were not.

This experience led to the decision to use rubber ring type joints for sizes 63 mm and larger so that contraction can take place at each joint without tension building up in the line. A new procedure was also adopted for solvent cement joints whereby the line is backfilled before 9 a.m. on the morning *after* the pipes have been joined. This practice has reduced the problem, even in some large pipe sizes which have had to be joined by solvent cement.

6.7 Steel Pipes

Galvanized steel pipes should only be used where mechanical strength is required. They are used at intakes; for crossing roads, rivers, and gullies; and for the pipework at tanks and standpipes. They are extremely expensive and heavy to transport. In Malawi sizes up to 100 mm were generally available, and larger sizes had to be specially ordered or welded locally.

6.8 Polyethylene Pipe

High-density polyethylene (HDPE) pipe is used in many countries as an alternative to PVC. The principal advantages are resistance to freezing, ease of laying with less joints than PVC, and greater flexibility. The principal disadvantages are higher cost, greater weight, and greater bulk for transporting. HDPE is supplied in coils which cannot be nested, so a lorry can transport relatively little HDPE pipe compared with PVC. This is significant when large quantities and larger sizes are involved. On the other hand, if pipe has to be carried long distances by porters in mountainous or remote areas, it is more efficient to carry coils of HDPE than lengths of PVC. PVC is used in Malawi because it is cheaper, internal transport costs are lower, and because there is a local PVC pipe manufacturer.

6.9 Technical Developments

It is significant that the programme in Malawi developed for 10 years without getting involved in more complex techniques such as impounding reservoirs, pumping mains, and water treatment. This is undoubtedly an important factor in the successful development of the programme.

The need for water treatment is likely to be the first sophistication that is considered. Ideally, rural water programmes should avoid having to treat water by choosing unpolluted sources and suitably protecting them. Eventually, however, polluted sources will have to be considered, and the most appropriate form of treatment in rural areas is likely to be slow sand filtration. The biggest problem with any sort of treatment is the need for competent operation and maintenance. It is unlikely that the community will be able to operate even a simple water treatment plant, which means that the government will have to supply trained personnel. This may be feasible for larger systems, but would be uneconomic for smaller systems.

Inexperienced programmes should therefore avoid the temptation of getting

Figure 6.7 Visitors from Ministry headquarters observe the jointing of steel pipes near an intake
(Photo: L.H. Robertson)

Figure 6.8 HDPE pipe coils are easier than PVC lengths to transport by porters in inaccessible terrain
(Photo: UNICEF, Nepal)

involved in more complex technologies until they have developed adequate technical capacity. The need for treatment or pumping should count as a major factor against the feasibility of a project. Otherwise, there is a danger that the scarce resources of technical expertise will become so absorbed in one particular technical operation that the rest of the programme will suffer from neglect.

6.10 Summary

Simple technologies require the same strict design criteria and attention to detail as do more complicated ones. Many problems, both technical and community-related, can be eliminated by detailed and accurate design. There is, however, a need to build in a degree of flexibility to cater for unknown factors that emerge at the construction stage. This is facilitated by combining the roles of design and construction in one engineer to enable feedback from field experience to influence design practice.

It is desirable to have reasonably detailed design procedures laid down in a simple form so that systems can be designed under supervision by sub-professional staff.

Relatively large-scale rural water supplies have been installed in Malawi involving very simple technology. The principal features of the gravity system are an intake, screening tank, and sedimentation tank, with a main line supplying branch lines via area and branch storage tanks. The system is designed for average flow, with the storage tanks balancing supply and demand. The Malawi experience has been favourable for both asbestos cement and PVC piping, as long as adequate supervision is guaranteed.

CHAPTER 7

Maintenance

7.1 Introduction

The maintenance of rural water supplies in developing countries has become the subject of increasing concern in recent years. Experience has shown that it is often far easier to construct a water supply than to ensure its continued operation. In this context much has been written about the advantages of community involvement and appropriate technology.

This chapter describes the experience of the Malawi programme in the field of maintenance and suggests a policy that may be applicable to programmes in other countries. It shows the development of a system in which the community can realistically play a role in maintenance, and the part that the programme must play to ensure success. It also describes what maintenance problems may be expected in a gravity system and how they may be minimized.

7.2 Factors Affecting Maintenance

Technology

The operation and maintenance requirements should be a major consideration when selecting or developing appropriate water supply technologies for rural areas. If the technology chosen is relatively complex, involving, for example, motorized pumps and water treatment plants, operation and maintenance in rural areas is likely to be a major problem. By comparison, some simple technologies such as untreated gravity piped water systems, or protected dug wells, have relatively low maintenance requirements. Other simple technologies, such as handpump systems, have a more significant maintenance requirement. No system, however, is maintenance-free, and careful provision has to be made even for the simplest system.

Standards of design, construction and materials

The maintenance required by a system is also directly affected by the quality of the design and installation work. As the programme develops, the maintenance load should progressively be reduced by closer attention to detail during design and by continuously improving the standard of installation. The combination of the roles of design and construction engineer ensures that designs are improved to reflect experience gained during construction. On the other hand, shortcomings in

Figure 7.1 Maintenance problems—this gully was eroded in 3 days of heavy rain. The pipeline had not been adequately protected

(Photo: the author)

construction are very difficult to rectify, which is why competent supervision is so important. The standard of construction refers particularly to intakes and tanks, the standards of pipelaying, the marking and protection of pipelines, and the construction of tap aprons. The quality of materials used, particularly pipes and fittings, will also have a significant effect on maintenance load.

Involvement of the community

The involvement of the community in the installation of its water supply is of major significance for subsequent maintenance. Undoubtedly, the greater the involvement, the greater the degree of responsibility felt by the community. *However*, this sense of responsibility must be supported in practical terms by a reliable maintenance organization.

The principal advantage of community involvement lies in the self-help labour available for maintenance and repair activities. The community is already familiar with the concept of providing self-help labour in return for technical support, and it is easy to maintain this relationship for maintenance purposes. It is, however, unrealistic to expect the community to take over *all* maintenance duties.

The second advantage arising from community involvement in the construction stage lies in the fact that the community is familiar both with the layout of the pipe network itself and with the technical operations involved in laying pipes. With this experience they are capable of carrying out minor repairs without supervision, provided materials are available.

Thirdly, the fact that the community feels a sense of responsibility greatly improves the surveillance of the supply, and ensures that defects are reported quickly. Some communities may be willing and able to employ someone to carry out routine maintenance work.

Technical and material support

The sense of responsibility shown by the community is not enough to *ensure* maintenance is carried out. The main resource the community has to offer is its labour, and possibly the ability to raise minimal sums of money. It is probably beyond both the human and financial resources of the community to construct and pay for a new intake that has been washed away in a flood, or even to obtain a single steel pipe to repair a wash-out at a gully. In addition, the community is unlikely on its own initiative to carry out all the routine preventive maintenance tasks that are necessary for the continued reliability and long life of the supply. It is therefore necessary for government to supply both the technical supervision and material support required.

7.3   Maintenance Policy in Malawi

The policy regarding maintenance has evolved along with all the other aspects of the programme. The initial policy proved to be too optimistic, and has been progressively replaced by a more realistic one.

Original policy

During the Pilot Phase the policy was to hand over completed projects to the community with sufficient spares and materials. This is a policy still applied in some countries. At the end of each project, the committee was instructed in the duties of maintenance, particularly in the repair of bursts and the cleaning of the intake, screens and tanks.

This policy did not achieve a sufficiently high standard of maintenance. It is true that the committees did repair bursts, and the intakes and screens were cleaned when they became blocked, but these were curative activities carried out because the supply had been interrupted. When the screens broke, or a pipeline was exposed by erosion, nothing would be done as long as the water continued to flow. Once the exposed pipeline was finally washed away, the committee would send a report to the programme office. This message would take a long time to get through and programme staff, who were busy elsewhere, inevitably would take a long time to react. Even then, the programme had to supply materials and carry out repairs with the resources available from the nearest project under construction, as there were no specific funds or staff available for maintenenace.

Attempt to involve district councils

The programme realized that large projects would need a more formal maintenance system. If, on a small project, the community failed to maintain its supply, only that community would suffer whereas, on a large project, the failure of one community would adversely affect all the others. In addition, the size of the system would mean that any interruption of the supply would cause the ingress of air and interfere with the supply for as much as 2 days after its repair.

The programme therefore decided to approach the Ministry of Local Government to persuade District Councils to take over the major supplies on completion. It was argued that the councils were already financially responsible for the maintenance of boreholes in the area which were now redundant as a result of the new supply, and the funds saved could be transferred to maintain the new gravity systems. The Ministry of Local Government initially agreed in principle, but the District Councils were naturally apprehensive, and were unwilling to take over until they were satisfied that they would be technically and financially capable of doing so, and that a viable maintenance system was already in operation. The project engineer in Mulanje therefore developed a maintenance system involving a Project Assistant with a specified annual maintenance schedule, a watchman to maintain the intake and headworks, and a generous supply of spares and materials, as well as a house and store. The only expenses the District Council was expected to bear were the wages of the Project Assistant and the watchman, and the cost of the transport of materials if necessary.

Despite progress at district level, the Ministry of Local Government became worried about the implications for all the other District Councils. Mulanje had a relatively wealthy and well-staffed council compared with others, and the Ministry felt that the maintenance of numerous water schemes throughout the country was basically beyond the resources of the majority of councils. The policy was therefore abandoned.

The involvement of a district council or its equivalent is clearly an attractive proposition in most developing countries. It absolves the central government of the burden of routine maintenance so that it can concentrate on the construction of more systems. However, it is probably true that most district level organizations are severely overburdened and have insufficient funds and technical capacity to provide adequate support.

Current policy

Under the current policy in Malawi, the programme itself retains overall responsibility for the water supplies it constructs. Most of the maintenance work is carried out by the community, but the programme provides a support framework in the form of staff and materials. Experienced Project Assistants are assigned for maintenance duties on the basis of one per approximately 40,000 consumers. His role is to monitor the state of the system and to train and motivate the community to carry out routine preventive and curative maintenance. Committees are established at different levels, each with specific maintenance duties. Adequate spares of pipes and fittings are allowed for in the original project budget. All recurrent costs of providing this framework are borne by the programme, except the cost of

Maintenance

replacement taps; this cost is borne by the village concerned. Although the main elements of the policy are already in operation, the institutional framework is still being developed.

7.4 An Appropriate Maintenance Policy

In developed countries, rural water supplies are run by public utilities and all the consumer has to do is pay the cost. In developing countries there is neither the trained manpower nor the infrastructure in rural areas for the government totally to manage and maintain rural water supplies. Even if this were possible, the consumers would not be able to pay the cost and no government could afford to run the supply without any revenue. The only possible solution is that the consumers should maintain the system themselves.

However, experience has shown that, in general, consumers are not able to maintain rural water systems. They often do not have the financial capacity to pay the cost of spares, they usually do not have the technical capacity to carry out anything but minor repairs and, most important, they usually do not have an understanding of what maintenance is. They may be willing and able to carry out minor *repairs*, but they are unlikely to provide the type of *preventive* maintenance that is necessary to prevent the whole supply rapidly deteriorating and eventually becoming unworkable within a few years.

It is clear, then, that an appropriate maintenance policy must take these two facts into account. Maintenance must therefore be a *shared responsibility*, utilizing the resources of the community to the full and involving the government to provide the necessary support and supervision.

The role of supporting agency

The role of the supporting agency is to monitor the state of the system, to encourage and motivate the community to carry out maintenance tasks, to ensure that spares are available, and to carry out major repairs beyond the capacity of the community. For this role, the agency must set up a maintenance organization with an appropriate budget and full-time staff. Whether the organization is run by central or local government will depend on the capacity of local government institutions. It is likely that local government will only be able to assume responsibility if it is supported from the centre in terms of budget, manpower, training, etc.

Maintenance offices and stores must be set up at local level and maintenance field staff, here called Maintenance Technicians, appointed to cover specific areas.

Maintenance Technicians

Maintenance Technicians should be selected from among the experienced field technicians of the rural water programme. As far as possible, the selected technicians should have worked in the particular area, and be familiar with some of the systems and the community leaders. Maintenance work carries high responsibilities, as it is likely to receive less supervision than construction work. Only the most reliable and

Table 7.1 Annual maintenance schedule for Mulanje West Water Supply, Malawi

Month	Lines	Duties
January	All main lines 1, 2, 3, 4, 5	Inspect intake lines, tank, fittings and all taps off main; operate all sluice valves, flush points and gate valves; check all air valves; clean all tanks; submit inspection report
February	Main lines 1 and 2; branch lines ABC (including line to Chisitu Tank)	Patrol on first Monday; inspect lines, tank, and taps; clean all tanks; operate all gate valves; submit inspection report
March	Main lines 3 and 5	Patrol on first Monday; inspect lines, tank, and taps; clean all tanks; operate all gate valves; submit inspection report
April	Main line 4; branch lines GH	Patrol on first Monday; inspect lines, tank, and taps; clean all tanks; operate all gate valves; submit inspection report
May	Main lines 1 and 2; branch lines JK	Patrol on first Monday; inspect lines, tank, and taps; clean all tanks; operate all gate valves; submit inspection report
June	Main lines 3 and 5; branch lines LMN (including line to Mbiza)	Patrol on first Monday; inspect lines, tank, and taps; clean all tanks; operate all gate valves; submit inspection report

experienced technicians should therefore be chosen. In some countries, maintenance duties may involve more travelling and hardship than construction work, and adequate allowances should be paid so that maintenance is not seen as an unattractive job. In order to keep Maintenance Technicians motivated and familiar with latest developments, they should periodically be transferred to the main construction programme, say every 5 years. They should be given a house, if necessary, with an adjoining store for spares and tools. They should be given a bicycle where appropriate, or preferably receive an allowance for using their own.

The Maintenance Technician's job is to carry out a continuous programme of inspection according to a specified schedule. An example of the annual maintenance schedule for a major supply (240 km of pipeline, 460 taps) in Malawi is given in Table 7.1. If a number of smaller systems are to be inspected, a similar schedule should be drawn up. The Maintenance Technician should carry out the inspection with a member of the local Maintenance Committee. He should personally ensure

Maintenance

Month	Lines	Duties
July	Main line 4; branch lines OPQR	Patrol on first Monday; inspect lines, tank, and taps; clean all tanks; operate all gate valves; submit inspection report
August	Main lines 1 and 2; branch lines STV	Patrol on first Monday; inspect lines, tank, and taps; clean all tanks; operate all gate valves; submit inspection report
September	All main lines	Inspect intake lines, tanks, fittings, and all taps off main; operate all sluice valves; flush points and gate valves; check all air valves; clean all tanks; submit inspection report
October	Main lines 3 and 5; branch line VWZ	Patrol on first Monday; inspect lines, tank, and taps; clean all tanks; operate all gate valves; submit inspection report
November	Main line 4; branch line XY	Patrol on first Monday; inspect lines, tank, and taps; clean all tanks; operate all gate valves; submit inspection report
December		Spare

that all minor repairs and preventive work are carried out while he is present, including routine maintenance procedures such as cleaning tanks, operating valves, and flushing pipelines.

He should particularly ensure that villages keep their tap and standpipe aprons in a good state of repair, and he should have the power to plug the standpipe of any village that is failing to fulfil its maintenance responsibilities. This power should be derived from the Maintenance Committee and should only be exercised after suitable warnings and consultations.

For each system the Maintenance Technician should have an aerial photograph or map showing the layout and the position of all valves, tanks, taps, etc. He should also have a check-list of all the points to be inspected, as well as a list of the names of the Maintenance Committee members. He should complete a report form for every visit to the system. This report will record any pipe bursts that have occurred and been repaired by the villagers as well as any spares that have been used.

The number of Maintenance Technicians required will vary according to the size of the systems and their location in relation to each other. As a guideline for calculating how many are needed, each part of every system should be inspected at least twice a year. In Malawi, where communications are relatively easy and the systems fairly large, the ratio of one Maintenance Technician to 40,000 is probably the lowest feasible; in some countries a ratio of one to 10,000 may be necessary.

In some cases, particularly on a major system where the headworks may be situated some distance away from the consumers, it is necessary to employ a watchman to carry out maintenance and surveillance duties at the headworks. His job is to keep the intake area clean, to clear the intake pipes and screens of debris particularly in the wet season, and to patrol the upper part of the pipeline which is often more vulnerable to erosion. The watchman need not be employed full-time— 1 or 2 hours work a day should be sufficient. He should be nominated by the Maintenance Committee and paid a suitable wage or in kind, either by the committee or by the supporting agency.

Supervision and stores system

The work of the Maintenance Technicians should be supervised by experienced Supervisors, who should regularly accompany them on inspection tours. The Supervisor should receive the maintenance reports and take appropriate action when necessary. Spare parts required for maintenance should be made available by the external supporting agency. Each Maintenance Technician should have his own stock of the most commonly needed spares, as well as a supply of tools. Maintenance Committees should also have adequate spares and tools to carry out minor repairs. A maintenance store should be situated at the Supervisor's office, which may be at regional or area level. A generous allowance for spares should be made in the original project budget, and extra spares and stores required must be paid for either by central or local government.

Cash contributions

In some countries it may be possible to raise money from the consumers, either on a per capita basis or on a consumption basis. Although this is clearly an attractive proposition for the external agency it should be treated with considerable caution as it often causes more problems than it solves. Consumers are usually reluctant to pay regularly towards a maintenance fund when everything appears to be working satisfactorily. The routine collection of money can become a major task which completely dominates actual maintenance work and can lead to a breakdown in the essential good relationship between the consumers and the water agency.

On the other hand, communities are usually able to collect some money for a very specific item, such as a replacement tap or parts for a handpump. These items are relatively cheap and the community can see that their replacement is necessary to restore their water supply. Such items can be made available on a cash basis at the maintenance store.

Role of the community

At the end of the construction of the project, the construction committee should be re-formed into a Maintenance Committee. In a large system, a hierarchy of Maintenance Committees is necessary at village, branch and main line levels. The duties of each Maintenance Committee must be specified by the external agency. This is an essential stage in the completion of a project. The duties are divided between those which are exclusively their own responsibility and those which they share with the Maintenance Technician.

Their own responsibilities should at least include the maintenance and repair of village standpipes, aprons, and drainage. In addition, the users of the land through which the pipelines pass should be responsible for ensuring the lines remain adequately marked and protected from erosion.

The responsibilities that the community shares with the Maintenance Technician include general surveillance, the maintenance of pipelines, cleaning of tanks, and the repair of bursts. In all these activities the committee is expected to organize the labour to work under the supervision of the Maintenance Technician.

Village maintenance worker

In the case where the Maintenance Technician is covering one or two major systems, or a clustered group of minor systems, he should be able to reach any trouble spot within a day or so. If, however, he has to cover a number of minor systems spread over a large area, it may take days before he receives a message and even longer for him to reach the trouble spot. In these conditions it is desirable to train a member of the Maintenance Committee to carry out some of the maintenance and repair functions. This Village Maintenance Worker should be identified during the construction of the project by the construction committee. He should then work alongside the field technician during construction so that he can become familiar with the layout of the system and learn the skills required to repair or replace pipes and fittings. On completion of the project he should be given a detailed maintenance routine which may involve 1 or 2 hours' work a day. He may be willing to do this unpaid, in exchange for the status of being responsible for the system, with the authority of the committee. In some cases, the community may be able to pay him an appropriate amount per month in cash or kind.

7.5 Maintenance and Repair Load

To give an indication of what maintenance is required in a gravity piped water system, a description is given here of the maintenance activities and repairs found necessary in the Malawi programme.

Intakes and screens

Apart from the routine cleaning of intakes and screens, repairs are periodically needed. The intake itself is the most vulnerable point of the supply and needs to be carefully sited, well-designed, and strongly built to minimize the effect of a flood. This lesson was learnt when, on at least three occasions, intake weirs were washed away or broken by flood water.

The galvanized steel intake pipes are subject to corrosion, especially as they are in a region of aerated water. The intake pipes on one supply needed to be replaced after 10 years of use. The programme now expects that all intake pipes will have to be renewed every 10 years—a fact that had not been anticipated. This replacement can represent quite a major expense on a large supply.

The screens in the screening tank also need periodic repair and replacement. If the screens are allowed to become completely blocked the wooden frames may break as pressure builds up on one side. The galvanized wire mesh also corrodes and breaks under pressure. The frames can be repaired as necessary, but the mesh probably has a life of about 5 years.

Tanks

Concrete tanks require little maintenance work apart from routine cleaning. Some tanks develop slight leaks, but these are not considered a major problem, and do not appear to get worse. The maintenance organization should include the capability of carrying out minor repairs to concrete works. Float valves connected at the inlets of tanks need to be inspected and may occasionally require the replacement of a washer.

General maintenance of pipelines

The pipeline itself is especially vulnerable to erosion during the first two wet seasons after installation. During this period the backfill and ridge over the line is less compacted than the surrounding earth, and surface rainwater will quickly expose the weak points in the line. Checkdams are built at gully crossings after backfilling, but these may be washed away or new gullies may be created. Thus there is a need for regular inspection of the line and *swift* action to repair or build checkdams. A further erosion hazard occurs where the pipeline crosses the ridges and furrows of cultivated fields. Unless taught otherwise, farmers tend to stop their ridges just short of the pipeline ridge, apparently so as not to interfere with it. Unfortunately, during heavy rainstorms, surface water runs along the furrows until it reaches the pipeline ridge, and then is free to flow downhill beside the pipeline ridge. The accumulated water from successive furrows soon turns into a stream which can seriously erode the pipeline.

Farmers are therefore instructed to *connect* their ridges to the pipeline ridge so that rainwater is trapped within each furrow. Ideally this should be done by self-help labour at the time of backfilling, but it still needs regular inspection to ensure that farmers are continuing the practice.

Pipelines are planted with drought-resistant *Paspalum* grass which should eventually dominate other vegetation. The purpose of the grass is to strengthen the ridge from erosion and to act as a marker.

The use of this grass has been only partially successful. The planting of grass along hundreds of kilometres of pipeline is a major logistical exercise. The operation usually occurs towards the end of a project when most villages have already received water, and project staff are very busy finishing off other jobs, and it does not get the attention it deserves. Additional methods of marking should be introduced, at least for main lines, such as concrete posts spaced at regular intervals.

Pipelines often become paths, a practice which is to be encouraged as it has the double advantage of effective marking and regular patrolling. Main pipelines are fitted with flushing points at low spots in the ground profile to flush out any sediment that may build up. These are opened twice a year by the Maintenance Technician, as are all sluice valves, to prevent the spindles from seizing. Air valves are also checked for leakage which may be due to a worn ball or to debris being caught after a burst has been repaired.

Asbestos cement pipes

Experience with completed supplies shows that once the lines have successfully withstood the test pressure, subsequent bursts are very rare indeed. Only one burst has occurred in 4 years in 29 km of asbestos cement pipe on Mulanje West supply, and that was due to mechanical damage following erosion. A very small number of bursts occurred after the testing of Phalombe supply, some of which were thought to be due to surge pressures set up by the excessively rapid operation of a sluice valve.

When a burst occurs on the asbestos cement main, the pressure drop is felt all over the first pressure stage. As soon as the Maintenance Technician notices the pressure drop, or is told about the burst, he cycles back along the pipeline until he finds the spot with the help of the people nearby. He then closes the nearest upstream sluice valve and opens the nearest flush point to reduce the erosion caused by the escaping water. He then instructs the nearest village headman to organize people to expose the pipeline for one pipe length either side of the burst. He also arranges for two people to collect a spare pipe from the nearest dump that has been specially left for this purpose. The Maintenance Technician then returns to fetch the tools and spare joints needed to effect the repair, all of which he can carry on his bicycle. The repair is effected by cutting out the damaged section with a hacksaw, and connecting a new section using short collar repair joints. The line is then flushed out and the trench backfilled. The supply should be restored within 12 hours of the burst.

PVC pipes

Rather more bursts are experienced with PVC pipelines than with asbestos cement in proportion to the lengths laid. There may be several reasons for this. PVC lines are not tested beyond normal static pressure, so many of the weak pipes and joints remain undetected. These are exposed later, possibly when surge pressures occur. In addition, PVC lines are buried at a shallower depth than asbestos cement and are more vulnerable to erosion and mechanical damage. PVC bursts may also be caused by poor laying procedures, especially the use of excessive solvent cement (which softens the pipe) and the inadequate preparation and backfilling of the trench. It is believed that PVC pipes may vibrate slightly when water flows under pressure, and a small stone may slowly wear a hole in the pipe.

As an example of the repair load on a major supply, 23 bursts and leakages were reported in 7 months (in the dry season) in an overall length of 208 km of PVC pipeline. Other bursts may have been repaired without being reported. This is a surprisingly high rate in comparison with previous experience, and it is likely that the average will be seen to be considerably lower once records are available from a number of supplies.

One of the advantages of PVC pipe is the ease with which repairs can be effected by the villagers themselves. Towards the end of a project, each committee is asked to select a villager for an afternoon of training at the project headquarters. The villager is taught how to make a solvent cement joint on PVC pipe, which he usually knows already, and how to make a socket on the end of a piece of pipe. The socket is made by holding the end of the pipe in a tin of boiling oil (old engine oil) for a few minutes until soft, and then pushing the hot end over the cold end of another similar size piece of pipe, and twisting the pipes against each other as the end cools. Once cool, the pipes are pulled apart, the oil is cleaned off and the socket is then ready to be used for a solvent cement joint. The villager prepares a number of sections of pipe with sockets on each end to serve as spares for effecting repairs, and takes them home together with some cleaning fluid, solvent cement, and a small brush for applying the cement. The only other piece of equipment needed is a saw to cut the damaged section of pipe, but this is a fairly common tool in the villages and can easily be borrowed when required.

When a burst occurs, the nearest trained villager is summoned to carry out the repair. He may obtain more materials from the Maintenance Technician when his supply runs short.

Taps

The tap itself is the most heavily used item in the whole water supply system. The programme uses ½-inch crutch-headed brass bibcocks with threaded spindles. Spring-loaded self-closing taps are not necessary as the people are used to treating water as a scarce and precious commodity and do not waste water which they themselves have worked for. However, on a major water supply with over 400 taps, over 100 taps have to be replaced each year, which implies an average working life of about 4 years. Replacement taps are supplied by the Maintenance Technician on receipt of the necessary payment collected by the village.

The most common problem stems from the failure to replace a worn washer, resulting in the user screwing down the handle until the drip stops. Eventually, the thread of the spindle becomes stripped and the tap is useless. A spare washer is issued with every tap, but it is not always fitted in time.

The second most common problem is a fault of the tap itself, namely that the handle becomes loose on the spindle and wears away the stem head until it no longer grips. This problem has been reduced since the manufacturer was asked to pin the handle to the stem.

The standpipe itself is unlikely to suffer any damage due to normal wear and tear. Very occasionally, however, one is maliciously damaged by a jealous or mad villager. In such cases the village must report the matter to the police and must pay the cost of a new standpipe, which is supplied and connected by the Maintenance Technician.

Maintenance

Aprons, drains, and soakaways

Although the construction of an apron, drain, and soakaway with each tap was seen to be a major step forward, there is still room for improvement in the quality of both the design and construction of aprons and drains. A high proportion of aprons develop minor cracks with a few years, which need to be repaired to prevent further deterioration. The responsibility for repair lies with the villagers themselves but, unlike the replacement of taps, there is little incentive for the community to repair an apron until the damage is intolerable.

At present, the apron is constructed by the villagers under the supervision of the Project Assistant who, in the absence of any local artisans, carries out the concrete work himself. All Project Assistants receive instruction in concrete work from the project builder, but this procedure does not ensure a uniformly high standard.

If the villagers are to continue to construct the apron, the programme should increase its technical support. The apron must be strongly designed, to allow for limited quality workmanship; Project Assistants should have more effective training in concrete work; and the supply of cement and chipped stone to each apron should be increased. This will put a greater load on the project, but it should reduce the subsequent maintenance load on the community.

An alternative would be to develop a small team of contract builders specializing in aprons, similar to the system for the construction of tanks. This would lead to a higher and more uniform quality of work. It would, however, create a greater logistical problem when a large number of aprons have to be constructed within a relatively short period of time; it would also reduce the villagers' sense of responsibility for the apron, making subsequent maintenance more difficult.

Soakaway pits are a relatively minor problem, requiring to be emptied and refilled with stones every 2 or 3 years.

7.6 Monitoring and Evaluation

Monitoring system

In the early stages of development, when the programme is small and closely supervised, the monitoring process can be carried out automatically by field staff. As the programme develops and expands, there is a need to institute a formal monitoring system. This should focus on the technical functioning of the water supply and be a part of the routine maintenance process. The Maintenance Assistant should send in monthly reports which should show the number of pipe bursts, taps replaced, tapstand aprons needing repair, tanks leaking, and how often the water supply has been interrupted. This routine observation should be fed back to the programme headquarters so that appropriate design and construction improvements can be made.

Evaluation

It is more difficult to carry out evaluation on a routine basis, although routine monitoring will play an important part. Evaluation tries to measure the level of service provided, the benefits, and the costs. Although routine monitoring of the programme plays an important role, evaluation requires a more area-specific and detailed approach. To acquire data for evaluation a few communities should be studied in depth, to measure water consumption; water use; water collection patterns; and the economic, social, and health benefits. The measurement of benefits is particularly difficult.

Health data are likely to be unreliable in rural areas, even if there is a clinic nearby. The diagnoses may be wrong, and many sick people never go to a clinic. A possible alternative method could be to compare the *nutritional* status of children inside and outside a water supply area, on the premise that exposure to water-related diseases will affect a child's nutritional status. The advantages of this method are that nutritional status can be relatively easily and cheaply measured, and it does not require skilled personnel. It should therefore be possible to get a large volume of data covering a high proportion of the child population. If this method shows a significant correlation between water supply, sanitation, and child nutrition, it could be developed to indicate the comparative health impact of different levels of service.

The details of monitoring and evaluation methods are not covered in this book as this aspect of the programme is currently being developed in Malawi. Feachem *et al.* (1978: *Water, Health and Development* (Tri-Med Books, London)) describe an in-depth, "one-off" evaluation of a water supply programme which could be used as a basis for the design of routine monitoring and evaluation.

CHAPTER 8

Benefits

8.1 Introduction

Water is a basic human need. We do not need to measure benefits in order to justify the provision of water *per se*, just as we do not usually quantify the benefits of food, shelter, or clothing. It may be argued that it is necessary to measure benefits to provide some economic justification for the enormous human and financial investments being made in water supply in developing countries. However, this argument does not seem to be applied to the far greater investments in water supply that have been made in developed countries over the past hundred years. The issue of benefits, therefore, should not be seen merely in terms of economic justification. Rather, it should be seen in terms of need to maximize the potential benefits of a water supply, which may or may not require actual measurement.

There has never been an external evaluation of the programme in Malawi to indicate what benefits have been achieved. The benefits described below are those perceived by the author based on over 10 years of experience in rural water supply programmes. The analysis is inevitably subjective, as there are virtually no data on which to base a scientific analysis. The brief discussion on the improvement of health benefits is based on the currently accepted academic viewpoint which makes sense to the public-health oriented engineer.

8.2 Benefits of the Malawi Programme

Social benefit

From the point of view of the consumers, the most significant benefit has been the reduction in time spent drawing water. On average, the consumer now has to walk less than 0.5 km to draw water, compared with anything up to 8 km previously. Life in rural areas of most developing countries is hard work, especially for women, and the provision of a water supply is probably the greatest single measure that can reduce the drudgery of life for rural women. The time saved may be spent on more productive activities, such as agriculture, or it may be absorbed by other domestic work. There may be a slight increase in leisure time, which must be counted as an important social benefit. The conventional approach is that women need special programmes to help them utilize the time saved in the most productive manner. This assumes that the women do not know what to do with their time, whereas in fact they are usually still extremely busy and may be reluctant to undertake new activities. This is not to deny the importance of programmes that help rural women lead a more productive and satisfying life, but significant social benefit may be realized even without such programmes.

Economic benefit

A rural water supply alone is unlikely to have significant economic benefit. However, it is often an essential part of the infrastructure necessary for rural development, and thus the economic benefits of rural development can be partly attributed to that water supply. One of the water projects in Malawi was part of an integrated agricultural development project. The area was fertile, but parts were under-populated due to the lack of domestic water. Over the years some villages had been abandoned due to the deteriorating water situation, and the inhabitants had migrated to the few remaining water sources, putting extra pressure on the surrounding land. The water supply enabled people to return to their original villages and thereby increase the area of land under cultivation. This resettlement began during the construction of the water supply.

Health benefit

There has so far been no attempt to identify what health benefits have accrued as a result of the rural water supply programme in Malawi. One reason is that the programme has developed in response to the expressed needs of the people, who are principally motivated by the *social* benefits. Nevertheless, both the Government and the donors see improved health status as a major justification for investment in rural water supply, so there is a need both to maximize the potential health benefits and to attempt to measure them.

There is, however, real evidence that the water programme has had a major impact in an *epidemic* situation. Figure 8.1 shows the incidence of cholera in Mulanje District during an epidemic in Malawi in 1973–74. The water supplies shown are the combined areas of Mulanje West and Chambe, and the small Migowi supply to the north of the mountain. Phalombe supply was not yet completed. Population density is uniform in the western half of the map, the south-eastern quarter being Mulanje Mountain. It can be seen that isolated cholera cases within the water supply area do not appear to *spread* within the community in comparison with cases outside the area. The outbreak occurred in the wet season when surface water becomes heavily polluted. The villages outside the water supply area were using the streams shown on the map, as well as unprotected shallow wells.

Political benefit

The construction of a water supply is often the most dramatic development activity that has ever occurred in a village. At a time of rising rural expectations and political awareness, a rural water supply programme is one of the most impressive ways in which the government can bring benefits to the people. It is worth mentioning, however, that if the water supply is poorly constructed and breaks down, it can have serious *adverse* political consequences.

Benefits

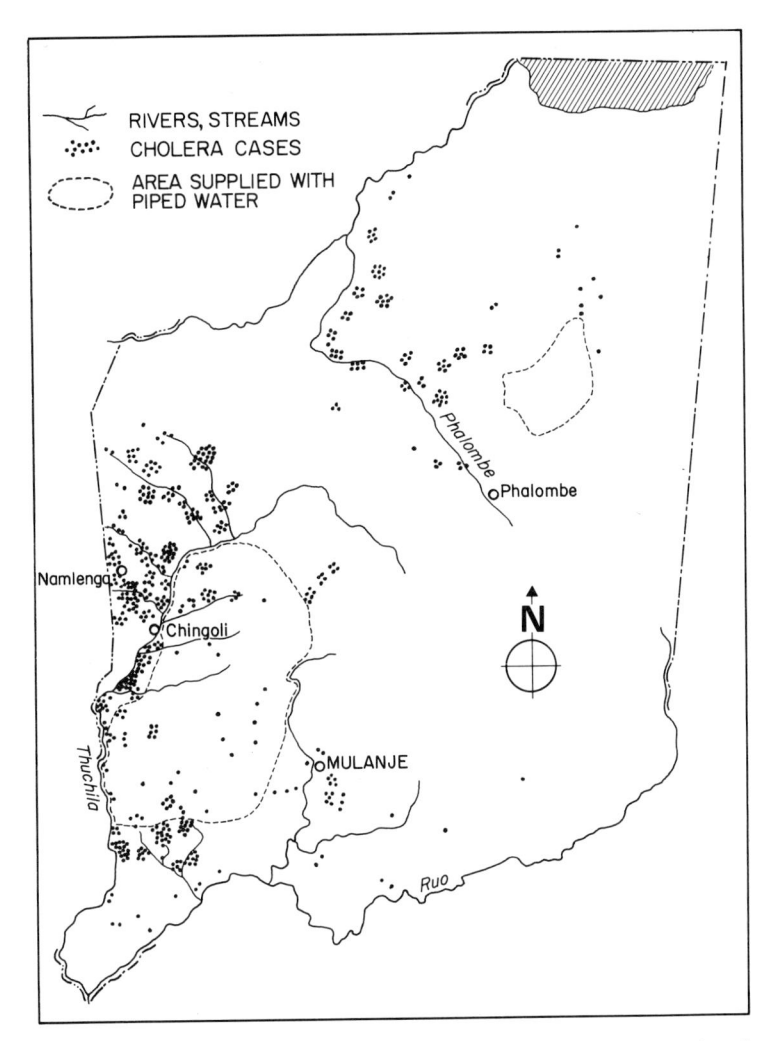

Figure 8.1 Incidence of cholera cases in relation to piped water supply, Mulanje District, Malawi, March 1974 Compiled by Dr E.C. Pauli and Mr L. Robertson and reproduced with their permission

8.3 Improving Health Benefits

Although it is difficult to measure health benefits, our knowledge of the transmission mechanisms of water-related diseases does indicate ways to maximize the potential health benefits (and minimize the potential health hazards) of a water supply.

Water use

Studies have generally failed to show a significant reduction in water-related diseases following the provision of clean water to a community. The logical conclusion is that transmission of these diseases continues by alternative routes related to *poor hygiene*. Standards of hygiene are likely to improve if a greater *quantity* of water is made available to the consumer, and if facilities are provided for people to wash themselves and their clothes. If people have to carry all their water home from a public tap, they are unlikely to use much more water than they used previously, even if the design allows them to. In Malawi, for example, per capita consumption on one system was estimated at 16 litres per day, although the design allowed for 25 litres. By providing clothes-washing facilities at the tapstand, as well as, say, a shower with suitable privacy for use by children and adults, transmission of water-washed diseases (relating to personal hygiene) is likely to decrease. However, there will be a significant increase in per capita consumption. The cost of the extra facilities, together with the cost of providing more water, may limit the coverage of the project. The designer has to decide the appropriate balance between benefit and coverage, which will vary from country to country, and even from project to project.

Sanitation

The majority of water-related diseases originate from the excreta of infected persons. No matter how much clean water is provided, transmission will continue as long as it is possible to come into contact with human wastes. For any significant reduction in these diseases it is essential that the community practises sanitary methods of excreta disposal. The water supply programme *must*, therefore, include a sanitation component if it is to have any significant health impact. This is the subject of the next chapter.

Health education

The decision of an individual villager to make more use of water, to improve personal hygiene, or to adopt sanitary methods of excreta disposal will be made for a complex variety of reasons—the desire to imitate, to conform, to be different, to gain influence, to improve social status, to achieve economic gain. Whatever the reasons, the decision must be preceded by some form of communication process. Health education is, therefore, a necessary but not sufficient condition for behavioural change. Although health education alone will not change people's habits, the introduction of a water supply can create a favourable climate for change, and it is essential that this opportunity is taken. Health education during the construction of a water system should be limited to a few specific messages relating to water use, personal hygiene, and sanitation, so that the potential health benefits can be realized.

CHAPTER 9
Sanitation

9.1 Introduction

The rural water supply programme in Malawi did not have a significant sanitation component. This chapter is based on the sanitation component that is currently being developed for the rural water supply programme in Nepal. The description below does not carry the same degree of first-hand experience as the rest of the book, as these developments have only taken place over the past two years. Nevertheless the approach is based on collective experience of working with communities in Malawi and Nepal and utilizes the same village-based, pragmatic approach that was successful in Malawi. Although it is too early to claim that this approach is successful, initial experience has shown that significant demand for sanitation can be stimulated in villages where sanitation is almost non-existent.

This chapter concentrates on the excreta-disposal aspect of sanitation. This is by far the most important, as human excreta is the most dangerous form of waste in rural areas. This is not to deny the importance of other aspects of sanitation; but, from a practical point of view, there is a far greater potential health benefit to be derived from concentrating on excreta-disposal, and making sure of success, than from spreading one's resources over a number of different activities such as solid-waste disposal and drainage, which can absorb the same amount of energy with relatively less health impact.

9.2 Strategy

Past experience suggests that public latrines are not suitable for rural areas. The arrangements for maintenance and cleaning such facilities generally break down. By contrast, programmes that have encouraged householders to build private latrines have been more successful. However, programmes that have forced people to build latrines—for example, by making a latrine an essential criterion to qualify for some other benefit—have generally failed. In addition, programmes that have provided latrines free of any cost to the people have generally not been successful on a large scale. It is essential that a villager builds a latrine only as a result of a conscious decision that he wants to *use* one. It is the use, rather than the construction, of the latrine that is crucial. The strategy to be adopted, therefore, is to encourage at least some villagers to decide that they want to *use* latrines, thereby stimulating a genuine *demand*.

The period of construction of a water system is an outstanding opportunity to stimulate such demand. The community is in a dynamic state—meetings are being held and discussion is taking place; field staff are regularly visiting, or may be resident in the community, giving the opportunity to bring in a few new ideas which are related to the water system.

9.3 Stimulating Demand

To stimulate demand for sanitation, it is necessary to create some awareness of the benefits. Very often, villagers are not aware of the link between human excreta and disease, and this relationship must continually be stressed. However, the villagers may be more attracted by other potential benefits of a latrine, particularly convenience, privacy, and status; these benefits should also be highlighted in the awareness campaign. The process of creating awareness takes a long time and a whole community cannot be expected to change lifelong habits during the period of construction of a water supply project. Nevertheless, *some* villagers *will* become aware during this period, and they are the seed from which the adoption process grows. The purpose of the awareness campaign is to create a favourable *atmosphere of reinforcing messages* for the seed to take root and grow. The awareness campaign should start right from the beginning of the water supply project.

Application form

The application form for the water supply, which should be completed in the community, should contain at least two questions about sanitation. The first question may ask "How many people in the community use latrines?", with multiple-choice answers "all: half: a few: none". The second question may ask "Does the community want help to construct latrines?" The answers to the questions are not particularly important; their purpose is to establish a link between water supply and sanitation (even if the link is not yet understood) and to generate some discussion about latrines among the leaders. Even if there is no discussion, there will be some subconscious impact of the question which will be reinforced by the next message.

Feasibility survey

The next opportunity to convey messages is during the feasibility survey. At the first public meeting (see Section 3.4), the team should stress that the programme is concerned with both water supply and sanitation. During the survey itself, the team should stimulate discussion about the relative purity of alternative sources and the need to protect from contamination whichever source is chosen. In the final public meeting the villagers who participated in the survey should be asked to report the discussions on which source has the best-quality water and how it could be protected. The survey team should not belabour the point or adopt a 'preaching' approach. The issue of sanitation should be brought in as a natural consideration when discussing alternative sources. Although many villagers will not understand this at the time, at least the question 'What is sanitation?' will form in their minds, which makes them more receptive to the next message.

Construction committees

It is essential that the committees become involved in the sanitation promotion process as early as possible. The project staff do not have the time or the influence to promote sanitation throughout the community on an individual basis, and they should therefore concentrate their effort particularly on the committees, which will include a number of local leaders. The main focus of attention for both the project staff and the committee is, of course, to organize the construction of the water supply system. It is the duty of the field technician during the first few meetings to lead the discussion towards sanitation. There are a number of ways of doing this, which can be developed with experience. For example, one question that can be asked to move the discussion in the right direction is 'What benefits are expected from the water supply?' If a committee member mentions that the new source will give cleaner water than the old source, the question should be asked 'Why?' In this way, by asking appropriate questions, the field technician can lead the discussion round to the problem of *disease* in the community. The opportunity should then be taken to emphasize that the source of many of the diseases in the community is not water but *human excreta*. This is a difficult point to make, especially when there is no understanding of the germ theory of disease, and it is highly desirable that the field technician has some communications material, such as a flip-chart to explain the relationship. If it can be arranged, one of the most convincing ways of showing this relationship is to let people see worms in excreta under a microscope. The assistance of the local health post may be necessary to organize such a demonstration. The point of the flip-chart or demonstration should be that if the community perceives health as a potential benefit of the water supply, this will not be achieved unless something is done to dispose of excreta. By further leading questions, the discussion can be directed to the use of *latrines*.

The above example should not be taken as a rigid procedure to raise the subject of latrines. It is an example of the type of approach that should be used, to guide discussion rather than to teach 'facts'. The discussion process usually takes more than one meeting, and points need to be repeated several times. There should be no illusion that a single discussion and flip-chart or demonstration will convince everybody; they may not even convince anybody at all, but they are a part of the atmosphere of reinforcing messages that the project needs to create, and they lead the committee towards the next step in the process.

Once sufficient interest has been aroused in the committee, the field technician should suggest that one (or more) committee members be nominated to assume special responsibility for sanitation. This member should help the field technician distribute propaganda material, address public meetings, encourage individuals, and report progress back to the committee at each meeting. In this way the committee assumes responsibility and a leadership role for sanitation, rather than the whole effort being left to the project staff. The committee member should be one of the first to construct his own household latrine, and should become the local 'expert', so that after the water supply project is over, there is someone left in the community to help those who later decide to build a latrine.

Public campaign

In addition to the involvement of the committee, there is a need to create an atmosphere of reinforcing messages directed at the public, using whatever public media are appropriate. Posters should be displayed at prominent points in the project area, such as at local stores, markets, meeting places, tea shops, bus stops, and government buildings. The implementing ministry should also buy time on the national radio to broadcast sanitation messages. Special material may be developed, such as films, slide shows, flip-charts, plays, songs, etc., to show to the public in the project area. These materials should be developed by the health extension services with the collaboration of the implementing ministry. The public campaign may be conducted by the health extension service, but in many developing countries this service is understaffed and may not have the capacity to focus on project areas. In this case project staff, namely engineers, supervisors, and field technicians, should conduct the campaign, using the material developed. A brief training course is necessary to enable project staff to do this.

At all public meetings in the project area, whatever the subject, project staff should emphasize the need for latrine construction. Those who have constructed latrines should be publicly congratulated at such meetings.

Focus on schools

Special attention should be paid to *schools* in the project area, for two reasons. First, school children need to be taught personal hygiene and sanitary practices of excreta-disposal. School children are more easily influenced and can be more easily taught than adults, whose habits are already well-established. Secondly, the children can be used to convey messages to their parents, thus contributing to the atmosphere of reinforcing messages.

The implementing ministry should ask the Ministry of Education to develop appropriate teaching materials for use by teachers in schools in the project areas. This material should consist of a short series of simple lesson plans, together with some larger posters to illustrate key points. The construction of a water supply in the community is an important learning opportunity for the children. The lesson plans should suggest activities that the children can carry out, such as measuring the length of a pipe, the depth of a trench, the volume of a tank, or the flow of water. The lesson plans should pose certain questions, such as 'What will happen if a pipe bursts? Who will mend it?' Stories may be used to explain how diseases are transmitted, how dirty water can spread disease, how water can become polluted by excreta. Personal hygiene should be stressed and practised. One lesson plan should ask the pupils how they could clean up the *school*, and should lead towards the need for latrines. Another lesson plan should ask how they could clean up their *homes*, and lead towards the need for latrines at home.

The field technician should ensure that the teachers receive the teaching materials, and should encourage their use. He should also be prepared to help with some of the lessons, for example, by showing pupils around the project, providing examples of materials used in the project and giving answers to some of the questions raised in the lessons. When the issue of latrine construction at the school is raised, the technician should offer technical advice and material support. The programme should have standard designs for school latrines and should make budgetary provision for materials not available locally. However, the construction of latrines at every school should not be an *automatic* function of the project. Rather, the project should be prepared to provide help to those schools that *ask* for it, as a result of the public awareness campaign and the specific teaching materials. In cases where the schools are managed by local committees, these should be involved in any decision to build latrines at the schools.

It is best to build separate latrines for teachers and pupils. If this is not done, the teachers either do not use a latrine at all, which is a bad example, or they reserve one of the pupils' latrines for their own use by locking it, reducing the number of latrines available for the pupils. The school tap should be sited next to the latrines to make it easier to clean them, and to encourage the washing of hands after using the latrines. The pupils should be involved in the construction of the latrines, by helping to dig the pit and collect construction materials. The project should supply materials that are not available locally, such as cement, reinforcing bar, ventilation pipes, etc., according to the design chosen.

Focus on health posts

Special attention should also be paid to involve the health posts or clinics in the project area. The staff of the health posts are valuable extra manpower to propagate sanitation messages to the public, and can have particular influence on patients who are suffering from excreta-related diseases. The field technician should ensure that the health post receives a set of communications material, such as the flip-charts, posters, films, etc. that have been developed for public use. It is, of course, essential that the health post has at least one latrine and that the staff have latrines at their houses. Again, the programme should use a standard design for health post latrines and should be prepared to provide technical advice and material support if necessary.

The latrine for the use of staff and patients is virtually a public latrine, and therefore it is essential that it is very easily cleaned. This means that it should have a slab of concrete, ferrocement, wooden planks, or some such material which can be easily cleaned. However, such a latrine is too expensive for the average villager to construct, so it would not be a suitable demonstration latrine. Therefore, in addition, a demonstration *household* latrine should be constructed at the health post entirely out of local materials. Outpatients who receive sanitation education messages from the health post staff during treatment can then be shown a latrine that they can build themselves at home.

Demonstration latrines

Propaganda alone will not persuade many people to build latrines. On the other hand, if, in addition, they are able to *see* a latrine of a type which they can build themselves, they can much more easily be persuaded. It is essential, therefore, to build demonstration household latrines in the project area. The field technician and the committee members responsible for sanitation *must* build latrines for their own use. It would be hypocritical if they tried to promote the use of latrines without using one themselves. The field technician should concentrate initially on helping these committee members to build latrines to the standard design.

9.4 Adoption Process

Early adopters

The purpose of creating public awareness and constructing demonstration latrines is to stimulate demand. Once the demonstration latrines are built, the field technician and committee members should encourage individuals to follow their examples. Particular attention should be paid to those who are likely to be persuaded earlier than others. The rate at which others will follow depends on the nature of the community. The process described here does not usually produce dramatically quick results, but the results are more likely to be genuine and long-lasting. Experience so far shows that an adoption rate of 75 per cent of households is attainable, but even 5–10 per cent in the 1- or 2-year duration of a water supply project should be considered a success. It is important that the adoption process at least *takes hold* during the project period, even if the spread is very limited. If it is felt that the cost of even the simplest latrine is so high for the average householder that it is preventing the process of adoption, a subsidy may be necessary, but should gradually be reduced as demand is stimulated. It should be remembered that if people are able to build a house, they should be able to build a latrine *if* they want one badly enough.

Follow-up

The rate at which the adoption process continues after the water project is over is heavily dependent on the continuation of the awareness campaign. The water supply programme will no longer be able to exert such influence and much of the follow-up campaign will have to be left to the health extension services. However, the Maintenance Technician who visits each community at least twice a year should be used to bring in fresh propaganda materials for the public, for schools, and health posts. He should also exercise his influence over the Maintenance Committees and Village Maintenance Workers to encourage them to build latrines.

The Maintenance Technician must visit institutional latrines on his inspection tours, to take appropriate action if they are being misused, becoming full, or needing repair. He should also visit a proportion of household latrines so that he can report whether or not they are being used properly.

9.5 Technical Factors

This book does not attempt to go into the technical details of different types of latrines, as these are well covered by other publications. The fundamental criteria to be applied are simplicity, minimum cost, design life, ease of cleaning, and social acceptability.

As the latrine is to be constructed by the householder himself, it should clearly be as simple as possible. The simplest type is the simple pit latrine. Ventilated Improved Pit latrines are more desirable but may have to be ruled out on grounds of cost. Any latrine requiring the use of water for flushing is extremely vulnerable to blockage. Users invariably flush with too little water, or may only flush once a day,

Sanitation

especially if water has to be drawn from a standpost some distance away. The water seal may trap rubbish and stones that children often throw down latrines. On these grounds, water-seal latrines are generally inappropriate for rural communities where there has been no previous sanitation. The cost of household latrines, which should usually be borne entirely by the householder, should be easily affordable. The householder should dig the pit; collect the construction materials such as wood, bamboo, stones, and clay; and construct the superstructure. This will usually rule out the use of cement and ventilation pipe in household latrines.

In some countries, where communications and transport are relatively easy, it may be possible for the programme to supply certain materials, such as cement to make a squatting plate and essential items to make a ventilation pipe. This will undoubtedly improve the quality of the latrine enormously. Ideally, the extra materials should be bought from the programme by the householder, so the cost must be kept to a minimum. However, only the richer people will be willing to buy these materials, so this improved design should only be available as an option. The basic design promoted by the programme should not *require* the use of such materials, so that it can be built by the majority of householders. If it is felt that in richer communities the extra cost would not be an obstacle once sufficient demand for the improved design has been generated, it may be appropriate to provide an *initial* subsidy to encourage the spread of this type of latrine. However, as mentioned above, there should be no subsidy unless it is clearly possible to phase it out within a few years. Few governments would be willing or able to provide such a subsidy to every rural householder in the country.

Even though the cost is minimal, the householder invests considerable time and effort in the construction of a latrine and may not be so willing to repeat it after a short time, especially when there is no field technician to encourage him. The longer he and his family use a latrine, the more dependent upon it they become and the motivation to build a new latrine will grow with time. The design life of the latrine should, therefore, normally be a minimum of 3–5 years. This also holds true for school and health post latrines.

Latrine squatting plates inevitably become fouled, especially by children. If they are not cleaned the situation rapidly deteriorates until people prefer not to use the latrines. This can happen within weeks of construction. A household latrine is more likely to be kept clean than a public latrine, but if the squatting plate is difficult to clean it will become more and more foul. In a household latrine the squatting slab should be given a hard 'polished' surface with clay, as is commonly done inside houses. There should be stone or wooden footrests to reduce the danger of hookworm infestation.

Institutional latrines, such as at schools or health posts, pose a different problem. It will be more difficult to find someone willing to keep the squatting plate clean, so it is even more important that the task of cleaning is made easy. For this reason these latrines should have a concrete or ferro-cement slab with a smooth surface.

Some sanitation programmes have experienced difficulties with the social acceptability of various latrine designs. This problem has mainly occurred where the design has been too sophisticated, involving unfamiliar materials or colours, and where the design has more or less been forced on to the community. It is also associated with communal rather than with household latrines. If the latrine is kept to its simplest form—namely a hole in the ground, with a strong, easily cleaned squatting plate and a superstructure of the same materials and design as used in the local houses—and if the latrine is built by the householder himself, there is little likelihood of any serious problem.

9.6 Conclusion

Because of failures by engineers in the past, there is a trend towards the idea that it is essential to have the expertise of a social scientist in all sanitation work. This may be correct if the 'master-plan' approach is adopted and if the designing engineers have no real knowledge of the people. However, even with a social scientist, such an approach is doomed to failure unless the engineers and technicians have real *community-level* experience. No engineer or social scientist should be considered competent to design or implement a sanitation programme unless they have personal experience of *both* the social and technical issues at community level; if they do not have such experience then the first step is to acquire it by living in the communities and working in some related activity for a while. If this is done, there is no need for a social scientist, as many of the issues that are endlessly debated in the capital city become much more obvious and understandable at village level.

In addition, the village-based approach that is described above is less likely to run into social problems, because the emphasis is more on persuading people to *use* latrines, and the type of latrine is limited to one that they can build themselves. Millions of people have used latrines for decades which are little more than a hole in the ground, a cover with a squatting hole, and a superstructure for shelter and privacy. Such latrines work because the users understand the need for them and genuinely want to use them.

CHAPTER 10

Implications for the decade

10.1 Introduction

This book has examined in some detail the experience of the development of the rural water supply programme in Malawi. However, as stated in Chapter 1, the main purpose of this book is to indicate certain fundamental principles and guidelines which are applicable to water supply programmes in other developing countries. That some of the conditions in Malawi were easier than exist in many other countries certainly accounts partially for the greater degree of success achieved, but it would be blind and irrational to suggest that Malawi is a 'special case' and that none of the lessons are relevant elsewhere.

The most important principles and features of this book are emphasized in this final chapter because of their special significance for the success of the Decade. The first part of the chapter stresses the crucial importance of *field management* in any rural development programme and summarizes those features essential for success. The second part concerns the nature of *programme development*, which is generally ignored in development literature, yet which is absolutely fundamental to an understanding of how a programme can develop successfully.

10.2 Field Management

Governments usually do not realize the amount of management and support needed to conduct a rural water supply programme. Communities are expected to sustain unprecedented feats of communal labour for a considerable period of time on projects which, however simple they may seem, require unexpectedly high standards of construction, often involving unfamiliar technology. Field staff are expected to manage and supervise far more work than is humanly possible with relatively little training or supervision. The subject of field management has been discussed in detail in Chapters 4 and 5 and has recurred as a constant theme throughout this book.

Selection and training of field staff

If one factor determines the performance of a programme more than any other, it is the quality of its field staff. Programmes should be careful to select the appropriate type of person who feels "at home" in rural areas and who will be accepted and respected by village people. An interview is not a sufficiently rigorous selection procedure to screen out inappropriate candidates; selection should take place *after* an initial short training course. It is essential that successful candidates are given an adequate period of on-the-job training under close supervision to enable them to accumulate some experience in community work and to practise standard routines and procedures.

Supervisory and procedural framework

The majority of field staff in rural development programmes are usually relatively inexperienced. They can only be expected to fulfil their responsibilities if they are supported within a framework of adequate supervision and are able to follow standard procedures for as much of their work as possible. This is imperative if adequate technical standards are to be attained.

Engaging community participation

The participation of the community should never be taken for granted. Field staff must play a *positive* role to ensure that appropriate committees are established, and must work closely with them to ensure the proper management of the project. Field staff have no direct authority over the villagers—their role is to guide the *committees* in the exercise of *their* authority.

Management of self-help labour

Adequate organization and management at the point of work is essential for the efficient utilization of self-help labour and to ensure that technical standards are upheld. If self-help labour is left unsupervised, technical standards will inevitably fall and community enthusiasm will gradually fade away. Field staff should ensure that the committee is involved in the daily supervision of work and should also personally supervise as much of the work as possible.

10.3 The Four Phases of Programme Development

Returning to the analogy made in Chapter 1, a living organism must have the right *conditions* in which to grow; clearly these conditions may vary according to the stage of growth—a more mature organism can survive a harsher environment than a less developed one. Although the development of a programme is a continuous process, it can be divided into certain phases, each with its own objectives and each requiring certain conditions for optimum growth. The historical analysis in Chapter 2 of the development process in Malawi revealed three distinct phases; namely the Pilot, Consolidation, and Expansion Phases. To these should be added a fourth, the Maintenance Phase. Each of these is now considered in turn.

Pilot Phase

The principal objectives of the Pilot Phase can be summarized as follows:

(1) *To assess and stimulate genuine public demand.* The strength of any rural

development programme is heavily dependent upon the nature and intensity of this demand. It is important to differentiate between the institutionalized demand of committees or politicians and the *genuine* demand of the communities. The communities may need water badly, but will not necessarily participate in a self-help programme until they know what this involves. The strength of genuine demand can be assessed by conducting pilot projects and observing how much demand is stimulated in neighbouring communities.

(2) *To test practical features.* The experience of the Pilot Phase will reveal technical shortcomings which must be rectified at this stage rather than later.
(3) *To assess manpower requirements.* The experience of the Pilot Phase will indicate the type and quantity of manpower required for each project.
(4) *To provide training and experience.* The staff engaged in the Pilot Phase will form the basis for the future management of the programme. The experience they gain will be used to train future staff and to develop appropriate management procedures.
(5) *To gain the confidence of government.* If the programme is to be allocated adequate financial and institutional support it must first prove that it is a feasible programme which fulfils a genuine public need. If external financing is necessary, a successful Pilot Phase will also attract potential donors.

The Pilot Phase consists of a number of pilot projects. To achieve the above objectives it is logical to ensure that the conditions are as favourable as possible for the pilot projects to *succeed*. The most important conditions are as follows:

(1) *Suitability of project communities.* Many projects experience problems because the community turns out to be divided and the leadership poor. These problems are compounded if an improved water supply is not considered a priority by the community. Before pilot projects are undertaken feasibility surveys should be conducted to identify coherent communities with strong leadership and a genuine need for water.
(2) *Project size.* The success of a pilot project is dependent upon the maximum possible supervision of all activities and on exceptionally close contacts between project staff and the community. As both the numerical strength and the experience of field staff in the Pilot Phase is invariably limited, pilot projects should be kept as small as possible.
(3) *Technical simplicity.* In all pilot projects, field staff will undoubtedly experience problems of a community nature which will require all their experience and energy to solve. If they also have to devote time and energy to resolving technical difficulties, the chances of success are severely reduced. Furthermore, a technically complex project generally takes a longer time to construct and places a heavier burden on the community. The enthusiasm of the community, which is often rather unreliable in the Pilot Phase, is likely to evaporate before the project is completed.
(4) *The quality and experience of field staff.* A pilot project is a particularly testing time for field staff, who will have to cope with many unfamiliar situations. They should, therefore, be experienced enough in community work to be

able to *foresee* and avoid as many difficulties as possible, and be persevering enough to guide the project through the problems that inevitably arise.

(5) *Accessibility of the project area.* The success of all pilot projects depends heavily on the guaranteed supply of all materials and on the regular supervision and support of the controlling ministry. If projects are situated in remote areas, communications will be difficult, supplies jeopardized and the projects may receive inadequate support. Furthermore, the first and last objectives listed above depend upon the projects being adjacent to potentially responsive communities and being reasonably accessible to government officials, politicians, and potential donors.

When starting up a new programme there may be a temptation to carry out pilot projects involving a wide range of technical options in a number of different areas. The intention of this approach is to assess the suitability of different technical options with different types of communities, including those which are known to be difficult. Although this approach may seem logical in theory, in practice it can be disastrous—invariably there is insufficient experienced manpower to provide the necessary supervision; there is usually insufficient expertise accurately to assess and compare the results of such a wide range of projects; it will be difficult to assess whether a project fails because of inappropriate technology or bad management, and the inevitable failure of some projects will seriously undermine the credibility of the programme. If experimentation is necessary, it should be carried out on a very small scale which can be closely monitored. The range of technical options should be limited initially to, say, two or three.

While many rural water programmes have already developed beyond this stage, there are often new activities such as sanitation, or the introduction of a different technology, which should pass through the Pilot Phase of development. In some cases, major new approaches and procedures necessary to rectify serious problems should be introduced in a Pilot Phase within a current programme.

The Pilot Phase should continue until the objectives have been achieved; if a project fails, another should be conducted utilizing the lessons of the first. The duration will vary from situation to situation, but even if all goes well the Pilot Phase requires a minimum of 2 years.

Consolidation Phase

Even when the objectives of the Pilot Phase have been achieved, the programme is still not ready to expand into a countrywide activity. Many programmes have run into serious problems because they expanded too rapidly after completing a few pilot projects. All programmes must undergo a period of *consolidation* during which they develop the competence and capacity necessary before expanding. The principal objectives of this phase are as follows:

(1) *To develop and standardize routine procedures and techniques.* These are essential for the following reasons. First, the shortage of trained manpower in developing countries calls for the maximum use of the manpower available. While adequate supervision will always remain one of the most important conditions for success, the degree of supervision required is dramatically reduced if field technicians and artisans use well-proven

standardized procedures and techniques. Secondly, inexperienced staff can learn the essential details of a process more quickly if it is taught and practised as a standard routine, than if they are expected to learn in a haphazard way as they carry out their work. For example the procedures described in Section 4.5 are virtually essential to achieve a reasonable standard of work with self-help labour; it would take each trainee years to develop these on his own but they are easily learnt as standard procedures during the first year of training. Thirdly the collective experience of all field staff who practise the same routine procedures leads to continuous improvements which can be incorporated at each annual refresher course. Without such procedures the collective experience would be so disparate that it would be extremely difficult to improve the skills of field staff in the same way.

In cases where alternative approaches are necessary these should be taught as alternative standard routines and the Supervisor should instruct the field technician which alternative to use in the particular circumstances.

(2) *To finalize technical designs and standards of construction.* The experience of the Pilot Phase is generally not sufficient to define the standards of design and construction. This should be achieved during the Consolidation Phase, when the programme is still relatively small and concentrated. It is disastrous to expand the programme until such designs are well-proven, because the ensuing problems will pose an intolerable burden on Supervisors and will hinder the development and introduction of improved designs.

(3) *To develop a manpower base* of well-trained and motivated field staff who are fully conversant with standard procedures and who have the experience and self-confidence to work in new areas under less supervision. This is the most important objective of the Consolidation Phase.

To achieve these objectives it is important that the Consolidation Phase is conducted under the appropriate conditions. A relatively small number of projects should be undertaken, within the management capacity of the relatively few field staff who have acquired experience in the Pilot Phase. The projects should be clustered together to facilitate maximum supervision and ease of communication. As in the Pilot Phase, the communities involved should be relatively united under strong leadership and technically complex schemes should be avoided. As many trainees as can be supervised should be recruited, so that they have as long a period as possible in which to gain experience under close supervision. To allow enough time for the main characteristics of the programme to be developed and adequate field staff to be trained, the Consolidation Phase should last at least 3 years.

Expansion Phase

This is the period in which the programme expands from a localized activity into a nationwide programme. It is vital that a policy of cautious and *gradual* expansion is followed. The fundamental condition for the success of this phase is that the government should announce a *firm policy* for expansion. Without this, it is impossible to prevent haphazard expansion under the influence of political pressures.

The following policy is based on the experience of the programme in Malawi, which was successful in containing such pressures.

First, a few specific areas should be identified in which there is a high potential for the growth of the programme and in which conditions are particularly favourable for development. These areas become the *focal points* from which expansion will take place. The number of focal points should initially be limited to, say, one or two per region of the country.

Secondly, the most experienced and competent field staff should be transferred to these areas to conduct a few relatively small-scale demonstration projects to stimulate genuine popular demand as well as to develop some local experience. These projects should preferably be completed within 1 year.

Thirdly, while the demonstration projects are being constructed, adequate management, logistic, and administrative support should be established at each focal point in preparation for expansion.

Fourthly, on completion of the demonstration projects, a few larger-scale projects should be conducted in adjacent communities where demand has been stimulated.

Fifthly, as each project is completed, new projects should be undertaken in adjacent areas, spreading outwards from the focal point.

Sixthly, new focal points should be established as the capacity of the programme permits. Expansion continues in this way until the country is covered.

The rate at which expansion should take place will obviously vary from programme to programme. For comparative purposes, it is significant that the population served in each year in the Malawi programme increased from 25,000 in 1972 to 100,000 in 1980, which is an average increase of 19 per cent per annum. This rate is probably unusually high because the programme involved a few large-scale systems which are relatively economical in terms of manpower and supervision. For most programmes the growth rate for the Expansion Phase should lie somewhere between 10 and 20 per cent.

If, for political reasons or to achieve an ambitious target, a programme is forced to expand at a faster rate, it will inevitably encounter serious problems. It will be unable to maintain a satisfactory level of experienced field supervision; technical standards will fall, and projects will begin to fail. Not only will the programme fail to reach its target but the rate of implementation may actually *decrease*. This will inevitably lead to a general loss of confidence and reduced support for the programme.

Maintenance Phase

Although responsibility may be transferred to another ministry or to district level (see Chapter 7), maintenance of completed projects is a distinct phase of the development process. The objectives of the Maintenance Phase are first to ensure that completed systems receive routine maintenance and minor repairs, and secondly to ensure the rehabilitation of systems that, as a result of neglect or natural phenomena, require major repair. The most important condition for proper maintenance is that a firm policy is developed during the Consolidation Phase. If the programme expands before the development of such a policy, the problem of maintenance rapidly becomes unmanageable. For example, a programme which implements 10 projects in the first year of expansion and grows at 15 per cent per annum will accumulate over 200 completed projects in 10 years.

The Maintenance Phase is probably the most neglected phase of development. The problems of implementation of rural water projects tend to absorb all the creative energy of management and field staff. In any case, governments are reluctant to divert precious manpower and financial resources from construction to maintenance; this is extremely short-sighted as political pressure is soon reasserted when completed systems break down. As far as the Decade is concerned, the enormous investment being made to construct more water supplies is being jeopardized for the lack of a relatively small investment in maintenance. To avoid disastrous consequences, governments should ensure that a realistic maintenance policy is developed and implemented at least by the middle of the Decade, and donors should make this a condition for continued support. To encourage this development, donors should be prepared to support the costs of operation and maintenance provided there is a similar commitment from the recipient government.

10.4 Conclusion

In March 1977, the United Nations Water Conference declared the period 1981 to 1990 as the International Drinking Water Supply and Sanitation Decade with the target of clean water and sanitation for all by 1990. This has undoubtedly been a successful public relations exercise to attract more international attention to the problem and to encourage much greater investment by governments of developing countries, as well as by multilateral and bilateral donors. Many countries have adopted national targets for the Decade and are in the process of developing and expanding their programmes accordingly.

'Water and sanitation for all' has so far proved a successful rallying cry to mobilize and launch the Decade. But how far should this slogan be carried? This book has stressed the danger that continued adherence to Decade targets may force many national programmes to undertake more work than they are capable of carrying out satisfactorily. This will result in the proliferation of poorly constructed and badly maintained water and sanitation facilities and in the disillusion rather than the satisfaction of communities which are so desperately in need. The potential health benefits of the Decade will not be realized, and instead of the political benefits that many governments expect, political pressure will be stronger than ever before. 'Water and sanitation for all' will remain nothing but a slogan for generations to come.

It is significant that the programme in Malawi took approximately a decade to develop from inception into a strong programme. The *real* target for the Decade should be the development of strong, viable national programmes, staffed with experienced management and field personnel, following well-proven policies and procedures and enjoying the confidence of the communities to be served. If this target is achieved, the Decade will indeed have been a success and the goal of water and sanitation for all will at least become a possibility within the lifetime of today's children.

APPENDIX 1
Rural piped water projects in Malawi at January 1979

(a) Completed rural piped water projects in Malawi at January 1979

Project	District/Region		Design population	Length of piping (km)	Number of taps	Construction period	Cost of construction (Kwacha)	Funding agency
1. Chingale	Zomba	S	5,000	40	35	1968–69	6,000	USAID
2. Chambe	Mulanje	S	30,000	97	180	1969–70	64,000	OXFAM
3. Migowi	Mulanje	S	6,000	24	45	1969–71	12,000	USAID
4. Chilinga	Mulanje	S	2,000	10	14	1971–72	4,000	CSC (Christian Service Committee)
5. Ng'onga	Rumphi	N	2,000	18	20	1971–72	6,000	CSC
6. Muhuju	Rumphi	N	1,000	19	21	1972–73	7,000	USAID
7. Chinkwezule	Kasupe	S	700	2	7	1973–74	1,000	CSC
8. Ighembe	Karonga	N	4,000	18	36	1973–74	7,000	CSC
9. Mulanje West	Mulanje	S	75,000	238	460	1972–75	170,000	UNICEF
10. Luzi	Mzimba/Rumphi	N	8,000	60	42	1974–75	24,000	CSC/UNICEF
11. Chinunkha	Chitipa	N	4,000	26	51	1974–75	12,000	CSC/UNICEF
12. Chilumba	Karonga	N	4,000	18	29	1974–75	8,000	CSC/UNICEF
13. Chilobwe	Ntcheu	C	1,200	6	12	1974–75	2,000	CSC/UNICEF
14. Phalombe	Mulanje	S	90,000	491	578	1973–77	500,000	DANIDA (Denmark)
15. Dedza	Dedza	C	1,400	8	10	1975–76	5,000	CSC/UNICEF
16. Mchinji	Mchinji	C	20,000	137	116	1974–76	52,000	CSC/UNICEF
17. Chagwa	Kasupe	S	7,000	80	95	1975–76	15,000	CSC/UNICEF
18. Kalitsiro	Ntcheu	C	1,000	6	9	1976–77	3,000	CSC/UNICEF
19. Lifani	Zomba/Kasupe	S	20,000	101	140	1975–77	72,000	CSC/UNICEF
20. Hewe	Rumphi	N	8,000	42	42	1975–77	30,000	CSC/UNICEF
21. Nkhamanga	Rumphi	N	12,000	76	120	1976–78	134,000	CSC
22. Lizulu	Ntcheu	C	6,000	24	25	1976–78	20,000	CSC/UNICEF
Totals			308,300	1,541	2,087		1,154,600	

C = Central Region, N = Northern Region, S = Southern Region

(b) Current rural piped water projects in Malawi at January 1979

Project	District/Region		Design population	Length of piping (km)	Number of taps	Construction period	Cost of construction (Kwacha)	Funding agency
23. Ntonda	Ntcheu	C	25,000	121	140	1977–79	106,000	CSC/UNICEF
24. Lingamasa	Mangochi	S	12,000	43	48	1977–79	40,000	CSC/UNICEF
25. Namitambo	Chiradzulu/ Mulanje	S	50,000	257	350	1976–80	480,000	DANIDA
26. Sombani	Mulanje	S	40,000	185	300	1977–80	240,000	ICCO (Netherlands)
27. Zomba (Domasi)	Zomba	S	100,000	451	700	1977–80	520,000	CEBEMO (Netherlands)
28. Luwazi	Mzimba	N	8,000	80	54	1978–79	79,400	CIDA (Canada)
29. Mulanje South	Mulanje	S	45,000	293	394	1979–81	150,000	CIDA
30. Karonga	Karonga	N	30,000	196	250	1979–81	180,000	CIDA
Totals			310,000	1,626	2,236		1,944,400	

(c) New rural piped water projects in Malawi commencing 1979

Project	District/Region		Design population	Construction period	Estimated cost of construction (Kwacha)	Funding agency
31. Kawinga	Machinga	S	60,000	1979–82	711,300	DANIDA
32. Nthalire	Chitipa	N	3,000	1979–80	50,000	CIDA
33. Dombole	Ntcheu	C	16,000	1979–81	85,000	CIDA
34. Mwanza Valley	Chikwawa	S	20,000	1979–81	120,000	CIDA
35. Livulezi	Ntcheu	C	10,000	1979–81	75,000	CIDA
Totals			109,000		1,041,300	

1 Kwacha = 1.23 U.S. Dollars (1979)

APPENDIX 2

Job Descriptions of the Malawi Programme

1. Senior Water Engineer

(1) The Senior Water Engineer is responsible to the Chief Community Development Officer* for the overall management of the programme.
(2) He is based in Ministry Headquarters.
(3) He is responsible for the long-term planning of the orderly development of water resources to supply rural communities, in cooperation with other government agencies involved in the sector.
(4) He conducts an annual feasibility survey of all project requests received in the year, and submits recommendations of suitable projects to be included in the programme.
(5) He liaises closely with Development Division and the Ministry of Finance in negotiations with aid donors.
(6) He liaises with Civil Servants at the appropriate level of related Ministries.
(7) He is responsible for the overall manpower, recruitment, and training policy, to ensure adequate staff are available for the execution and expansion of the programme.
(8) He is responsible for procurement of all supplies,† including quotations, orders, shipping instructions, and payment.
(9) He is responsible for the overall transport programme.
(10) He is responsible for the detailed design, material, and staff requirements for all projects.‡
(11) He monitors the progress of all projects, making regular visits and maintaining a close relationship with field staff.
(12) He is responsible for the establishment and execution of an adequate maintenance policy for completed supplies.
(13) He is responsible for the establishment and review of design criteria.
(14) He ensures that Technical Officers receive appropriate training in design and fieldwork.

* With the formation of the new ministry for the water sector (1980), the Senior Water Engineer is responsible to the Chief Water Controller. Some of the duties listed will be taken over by a new post of Senior Projects Engineer.
† The procurement of locally available supplies is delegated to field officers.
‡ Design is delegated as much as possible to Project Managers and Technical Officers.

(15) He is responsible for all expenditure and accounts.*
(16) He conducts the development and adaptation of appropriate water supply technologies to expand the scope of the programme.†

2. Project Manager‡

(1) The Project Manager is responsible to the Senior Water Engineer for the overall management of the project.
(2) He carries out the detailed design and draws up the material and staff requirements for the whole project.
(3) He makes a schedule of project work for each year.
(4) He surveys and sites the intake, all tanks, river crossings, and the alignment of all main pipelines.
(5) He is responsible for the ordering, procurement, and distribution of all local supplies within the funds allocated, and for the receipt and distribution of all specially imported supplies.
(6) He is responsible, through the Supervisors and Project Assistants, for the motivation, organization, and supervision of self-help labour.
(7) He is responsible for overall supervision of field staff, ensuring efficient staff utilization, maintenance of technical standards, in-service training, and staff morale.
(8) He conducts regular staff meetings to monitor progress and problems.
(9) He supervises the work of the building contractor.
(10) He has control of all project vehicles and their maintenance.
(11) He is responsible for the final inspection of the project, to ensure the desired standard has been attained and to resolve any problems before the project completion date.
(12) He sets up the maintenance organization and ensures that it is functioning correctly before project completion.
(13) He submits simple quarterly progress reports.
(14) He liaises with the District Officers of all Ministries or Departments working in the project area.

3. Supervisor§

(1) The Supervisor is responsible to the Project Manager.¶
(2) He liaises closely with all committees and community leaders to explain the work to be done and to assist committees with the organization of the self-help work programme.

---

* In liaison with the accounts office of the Ministry.
† For example, SWE has developed a handpump for the Shallow Wells Programme.
‡ This job description is similar for Technical Officers in charge of smaller projects.
§ Under the new organization (1980) these are called Water Foremen to bring them into line with other staff in the water sector. They continue to have specialist skills and experience in community organization.
¶ There may be one or two Supervisors in a major project.

Appendix

(3) He is responsible for the day-to-day supervision of Project Assistants, monitoring their work programmes and progress reports.
(4) He pays particular attention to the support and advice for trainees.
(5) He is responsible for the day-to-day utilization of project vehicles.
(6) He is responsible that the project stores system is maintained efficiently and that all stores drawn are utilized correctly. He is assisted in this by a storeman.
(7) He is responsible for the efficient distribution and use of project tools.
(8) He is particularly responsible that all jobs listed on the Village Tap Check Sheet are completed before a tap is issued.

4. Project Assistant*

(1) He is responsible for a branch line or a section of a main line on a major project, either alone or as part of a small team of two or three.
(2) He is responsible for the day-to-day supervision and efficient utilization of self-help labour in close liaison with village headmen, committees, and leaders, and ensuring the labour achieves the standards required.
(3) He is responsible for marking the route of pipelines from aerial photographs.
(4) He is responsible for the operations of digging, laying, backfilling, protecting and marking his pipelines.
(5) He is responsible for the technical operations involved in joining asbestos cement, PVC, and steel pipes and fittings, and the construction and installation of village standpipes and aprons.
(6) He is responsible for the procurement of all pipes and fittings from the project stores in accordance with information interpreted from his aerial photograph.
(7) He submits a fortnightly work programme and a weekly report to his Supervisor.

---

* Under the new organization (1980) these are called Water Project Operators to bring them into line with other staff in the water sector. There are three grades according to experience and proficiency.

APPENDIX 3

Design procedure for gravity piped water systems in Malawi

1. Basic Design Criteria

(1) Consumption 27 litres per capita per day.
(2) Flow per tap 0.075 litres per second (originally 1 gallon per minute) assuming all taps in use at the same time.
(3) 160 persons per tap
(4) Night storage period 8 hours, day service period 16 hours

2. Materials Required for Design

Aerial photographs
Survey maps
Census maps
Census tables
Graph paper
Dividers
Coloured chinagraph pencils
Coloured felt-tip pens

3. List of Villages and Populations

This is drawn up using the census maps and tables.

4. Design Population

(1) *Present population*: this is calculated from the most recent census data applying a 2.6 per cent increase per year.
(2) *Potential population*: this will partly depend on the population that can be supported by the traditional agriculture of the area. The productivity of the soil is assessed in conjunction with the Ministry of Agriculture. The potential population will vary from about 100 to 300 persons per sq. km. It should be remembered that provision of domestic water will encourage settlement in uncultivated areas, thus increasing the population.

(3) *Project design population*: by interpreting the existing and potential populations, a project design population figure is chosen. This will determine the total quantity of water required from the source.
(4) *Design populations of villages*: the ratio of the design population to the project's present population will give a factor. The present population of each village is multiplied by this factor to give the design population of each village.

5. Marking the Aerial Photographs

Using the census maps and design population figures, the villages and populations are marked on the aerial photographs in blue chinagraph.

6. Tap Locations

Provisional tap locations are marked on the aerial photographs in red chinagraph. This is a technique that improves with practice and it is useful to study photographs of previous projects. Some important points are:

(1) When plotting tap locations reasonably definable areas of, say, 5 to 10 villages are considered at a time and the number of taps plotted according to the total design population of those villages. Villages are not considered individually at this stage.
(2) The *initial* allocation of taps is 1 per 180 persons. This means that approximately 10 per cent of the total number of taps possible are reserved to allow for omissions that may subsequently come to light. The *final* allocation is 1 tap per 160 persons, averaged over the whole project area.
(3) Taps are plotted at population centres for large communities and at mid-points between smaller communities. In sparsely populated areas of potential development, growth points are identified, e.g. at the intersection of paths. As long as the tap is near at least one or two houses, the rest of that village will help to dig the trench.
(4) In general, taps are kept away from the banks of rivers and streams.
(5) Every school and health centre is allocated a public tap.
(6) The final siting of the tap on the ground is left in the hands of the communities themselves, but intelligent siting at the design stage will minimize the changes necessary later.

7. Transferring Tap Locations onto Map

The tap locations are now transferred accurately from the photographs to the map.

Appendix

8. Alignment of Pipelines

The taps are now joined up on the map by a network of pipes, involving sub-branches, branches, and main lines, like the trunk and branches of a tree. The basic principles are:

(1) to maintain a steady gradient;
(2) to site the main lines and branch mains on a ridge to reduce soil erosion hazards;
(3) to keep the length of piping to a minimum by following the shortest route (unless this conflicts with (1) and (2) above).

The network is extended right back to the header tank. Several possible alignments are drawn to obtain the best. At this stage the alignment is drawn in pencil on the map.

9. Selection of Storage Tank Sites

This is done in conjunction with the alignment of pipelines. Tank sites are selected according to the following principles:

(1) a tank must be sited at such an elevation as to give an adequate gradient to the area it serves;
(2) as a tank marks the beginning and end of a pressure stage it must be situated so that the pipelines operate within the maximum working pressure of the class of piping used;
(3) tanks divide the supply into manageable areas, both for installation and operation—a large project may have a number of large area storage tanks, and ideally every branch line should have its own storage tank.

10. Storage Factor

The rate of inflow to a storage tank should be two-thirds of the outflow.

(1) Under design conditions, the outflow from a tank occurs for 16 hours per day, so the continuous inflow rate need be only two-thirds of the service time outflow requirement.
(2) For tanks in series, the storage factor is only applied for those taps served *directly* from the tank, not to taps served through a subsequent tank for which the storage factor has already been applied.
(3) The storage factor is never applied to a sedimentation tank which must be full all the time.

11. Flows Required

Once the tap locations, pipe alignments, and the tank sites have been plotted on the map, each section of pipeline is marked with the flow required in that section:

(1) Starting from the end tap on every branch, the flow in each section of pipe (i.e. between two connections) is marked on the map on the basis of 0.075 l/s per tap.
(2) At each tank the storage factor is applied where appropriate as described above.

12. Design of Main Line to Area or Branch Storage Tanks

(1) The lowest point of the main line is determined. This may be a tank at the end of the line if gradient is steady, or it may be at a river crossing anywhere along the line. The lowest point is determined (approximately) by checking the contours of the map.
(2) A suitable header tank elevation is determined by:
   (a) the need for an adequate overall gradient from header tank to the storage tanks;
   (b) the need for the maximum static pressure at the lowest point to be within the limit of the class of pipe chosen.
(3) A ground profile is plotted on graph paper of the main line from header tank downwards. This can be done from the map at this stage, prior to a full survey. The crossing of each contour should be plotted although some interpolation is possible using spot heights and streams.
(4) A horizontal static head line is drawn from the elevation of the header tank.
(5) The desired hydraulic gradient is lightly drawn in pencil. This is always above the ground profile and will either be a straight line from end to end, or a series of straight lines changing at high points in the ground profile.
(6) Knowing the flow required in the first section of main line (from Section 11 above) the pipe size is calculated that will give the required flow for a head loss nearest to the desired hydraulic gradient. The actual head loss in that section is then plotted.
(7) This is repeated for every section of the main line, plotting successive head losses, keeping the hydraulic gradient as close as possible to the desired hydraulic gradient.
(8) If the main line splits, the flows are kept in the right proportion by adjusting the elevations of the tanks at the end of each arm relative to each other. The hydraulic gradient of the two arms is kept approximately the same as that of the main line.
(9) It may be found that the maximum static pressure at the lowest point in the project is above the maximum working pressure of the class of pipe chosen. In this case three courses are open:

(a) If the maximum static pressure is not greater than 110 per cent of the maximum working pressure of the pipe, the excess is acceptable (pipes are always tested to double their maximum working pressure).
(b) If it is greater than 110 per cent, the next class of pipe should be used for the high-pressure section. However, for asbestos cement pipes this causes complications with fittings, and it is only considered if the high-pressure section is a small proportion of the main line, say less than 25 per cent. If it is longer than this, the whole main line of that particular size should be of the next class of pipe.
(c) An alternative is to keep the end of the main line open the whole time (i.e. flowing into a tank) so that it can never reach static pressure, and the maximum pressure is then the normal running pressure. This means there must be no sluice valves in the line.

(10) Maximum pipe weight for man-handling is 120 kg. The design must not call for a pipe size which is impossible to handle by self-help manpower. This should normally be considered a limiting factor on the size of a project.

13. Design of Branch Lines from Storage Tanks

(1) A ground profile of the branch line is plotted from the storage tank, showing every pipeline and every tap. This is done from the map and drawn on graph paper.
(2) The flow is written against each section of pipe between two connections.
(3) The pipe size is calculated for each section of the branch main, and the head loss plotted. Ideally the hydraulic gradient is kept roughly 10 m above ground level, to allow for contour errors on the map and to ensure every tap has a positive pressure. Calculations are made using nomograms or a water flow slide-rule calculator.
(4) The pipe sizes are calculated for all minor branches down to each tap. The normal size for a single tap connection is 20 mm (except high-pressure taps—see (9) below).
(5) The hydraulic gradient for every section of pipe is plotted in pencil on the profile so that the design running pressure at any point is known. This is important for solving flow problems that may be experienced after laying.
(6) When every pipe size has been calculated, the pipelines are coloured in on the profile using the colour code.
(7) Normal PVC pressure class used is class 10 (maximum working pressure 100 m head). For large pipe sizes, 110 mm and over, class 6 is sometimes chosen for reasons of economy.
(8) If pressures over 100 m are involved, a break-pressure tank is installed.
(9) All taps on a branch main of size 50 mm and larger, and all taps at which there is a plotted pressure of over 10 m, are marked on the profile with a red chinagraph circle. Sizes 16 mm and 12 mm are usually used to connect these high-pressure taps to ensure they do not take excessive flow.

14. Storage Tank Capacities

Under design conditions, the night storage period is 8 hours. The size of the tank must therefore be sufficient to store all the water supplied at the design flow rate in a period of 8 hours. The next highest standard size is chosen. Standard tank sizes at present used are 14, 27, 46, 68, 91, 137 and 228 m^3.*

15. Plotting Design onto Map and Aerial Photographs

(1) Once the design of pipe sizes has been completed, the pipelines in pencil on the map can now be overdrawn using the colour code for different pipe sizes.
(2) The pipe sizes are marked in red figures alongside every section of pipe on the aerial photographs.
(3) Tanks, with sizes and elevations, are also marked on the map and photographs.
(4) High-pressure taps are transferred from the profile on to the map and photographs.

16. Siting of Air Valves

(1) Single and double air valves are designed to release air that is trapped in the water in the form of tiny bubbles, which accumulate at high points and can cause an airlock. Double air valves have the additional property of allowing air to exhaust rapidly from the line during filling, or preventing the build-up of a vacuum after a burst.
(2) The profile of the main line plotted after survey shows the approximate position of high spots. On a large main (size 100 mm and above) double air valves are situated approximately every 3 km at high spots and changes in gradient. Single air valves are fitted at other high spots. Approximate positions are marked on the profile, map, and photographs, the exact position being fixed later by surveying each spot in detail.

17. Siting of Flush Points

(1) The purpose of a flush point is to empty the main and to clean out any sediment that may collect at the low points. It consists of a tee-piece with 80 mm branch controlled by a sluice valve.
(2) The main line profile shows the position of low spots, usually at stream crossings or swamps. A flush point is fitted about every 3 km and at every significant low spot.

---

* Equivalent to 3, 6, 10, 15, 20, 30 and 50 thousand gallons.

Appendix 143

18. Siting of Sluice Valves and Gate Valves

Sluice valves are used for sizes 80 mm and above, gate valves for sizes 100 mm and below. Valves are sited as follows:

(1) *At the inlet and outlet of every tank*, except at the outlet of the main header tank (to prevent surge pressures caused by sudden closing or opening the valve. The main line can be closed by emptying the tank).
(2) *In main lines* at every branch and every change of pipe size.
(3) *In branch lines*:
   (a) at the beginning of every branch where the major line is 40 mm or larger;
   (b) at every reduction of pipe size involving 40 mm or larger;
   (c) below 40 mm at every size reduction or every 1.5 km, whichever is the less.

Valves are marked on maps and photographs with the appropriate symbol.

19. Design of Pipelines at Headworks

(1) Having decided on sites for intake and screening tank, the pipe alignment from intake through the screening tank to the sedimentation tank is selected, surveyed, and plotted on the profile.
(2) The inflow to the sedimentation tank is designed to equal the outflow, plus about 10 per cent to ensure a continuous overflow.
(3) Knowing the flows and the head available, the pipe sizes can be calculated.

20. Calculation of Material Requirements

Once the design is completed and plotted on both map and aerial photographs a list is drawn up of pipes, fittings and construction materials. This is used for estimating costs and ordering materials.

(1) *Asbestos cement (AC) pipes*: The lengths are measured from the map and 3 per cent is added for errors and breakages.
(2) *Cast iron fittings for AC pipe*: A schedule of fittings is made out showing the figures derived from "actual count" plus an allowance for spares. The most common spares needed are short collar joints and a few extra saddle pieces. One or two extra hydrant tee pieces are included for unforeseen air valves or flush points.
(3) *Major flow control fittings*: Sluice valves, air valves (double and single), and equilibrium float valves are counted and an allowance made for spares.
(4) *PVC pipes*: The lengths required of each pipe size are added *accurately* from the map design: 3 per cent is added for sizes 40 mm and larger, 5 per cent for 32 mm, 10 per cent for 25 mm, 20 per cent for 20 mm, and 10 per cent for 16 mm.

(5) *PVC pipe connection fittings*: The fittings required on each line are counted and a sensible allowance is added for spares.
(6) *Gate valves, float valves, taps*: These are counted from the plan and a sensible allowance added for spares.
(7) *Steel pipes*: Intake pipe requirements are calculated in detail and estimates are made from the map design of the sizes and quantities needed for river, stream, and gully crossings.
(8) *Construction materials*: At this stage a list is made of the number and sizes of tanks and all other works (intake, river crossings) and the number of tap sites.

21. Presentation of Completed Design

The completed design consists of:

(1) A sheet showing:
 (a) actual population;
 (b) design population;
 (c) number of villages;
 (d) total land area;
 (e) estimated arable land;
 (f) total design daily consumption;
 (g) per capita daily consumption;
 (h) number of taps;
 (i) ratio of persons per tap;
 (j) total storage capacity.
(2) A list of material requirements.
(3) A diagram, not necessarily to scale, of the layout of main lines to area and branch storage tanks, showing pipe sizes, length, flow, and head loss in each section and the size and elevation of tanks.
(4) Profiles of main lines.
(5) Profiles of branch lines.
(6) Map showing design.
(7) Aerial photographs showing design
(8) List of villages and populations.
(9) All sheets used for calculations (useful if design changes have to be made).

APPENDIX 4
Example of Use of Topographical Maps in Malawi

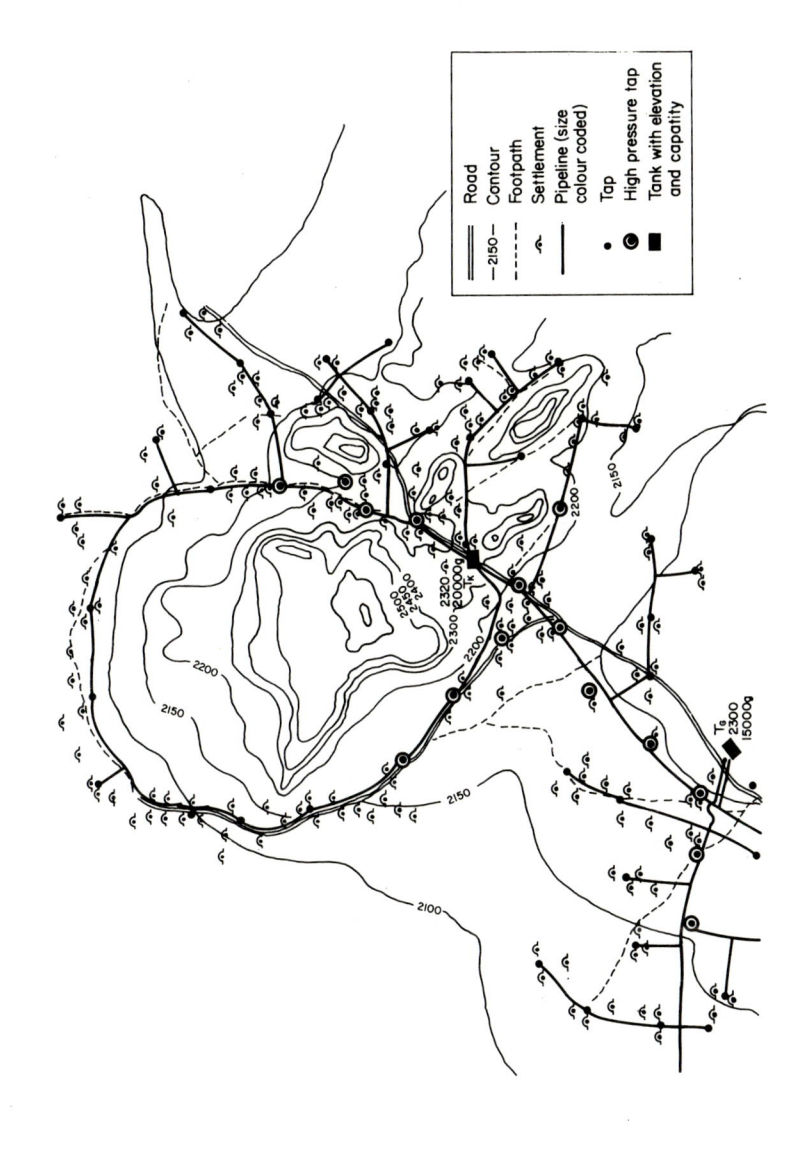

This simplified topographical map (original scale 1:50,000) shows the layout of part of Sombani Water Project in Malawi. The sites of tanks and taps are marked, as well as the routes of the pipelines, which are normally colour-coded to indicate pipe diameter.

APPENDIX 5

Example of Use of Aerial Photographs in Malawi
(original scale 1:20,000)

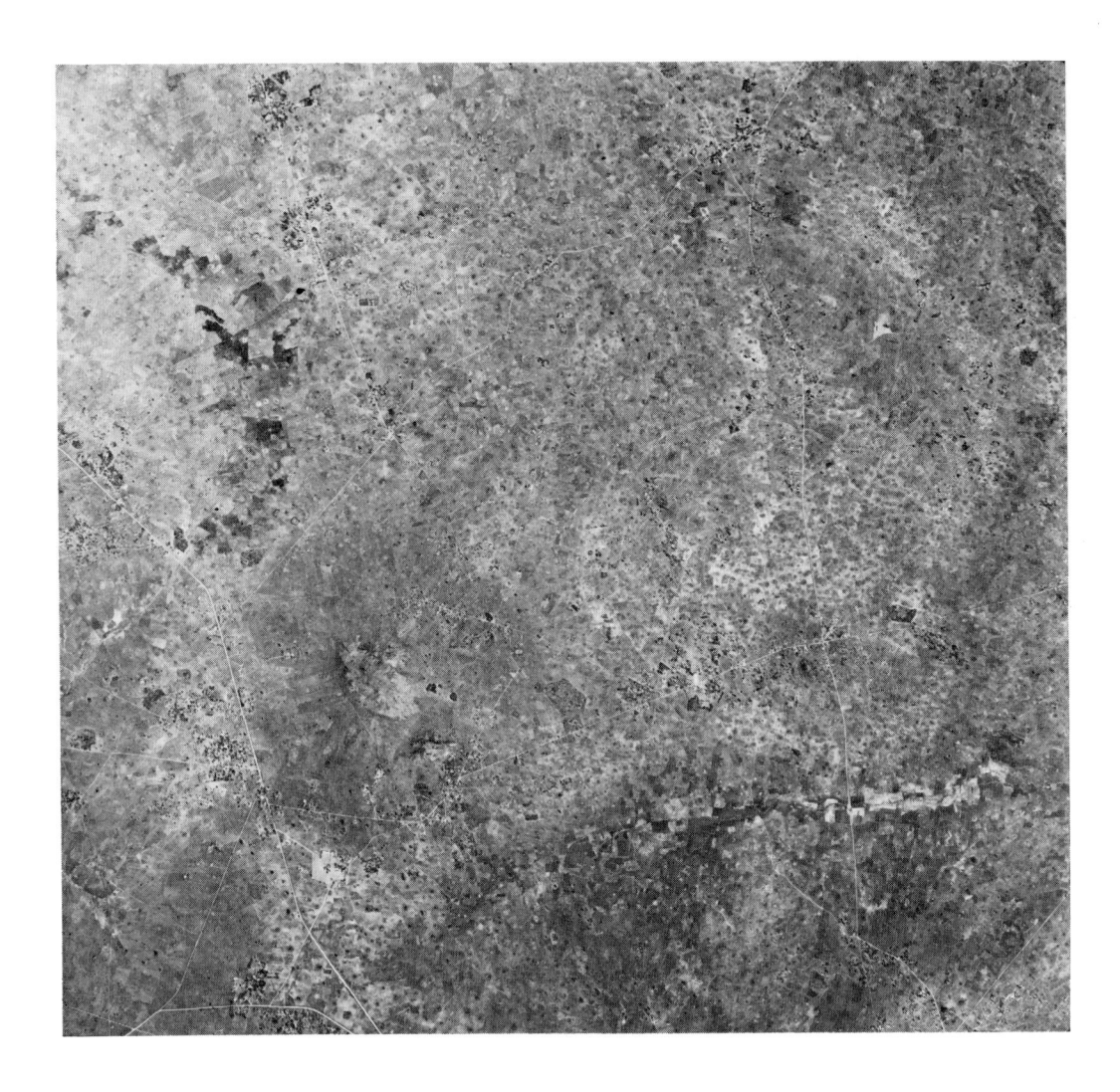

Aerial photographs can be marked up with coloured chinagraph pencils to show the features of the pipeline. For the sake of clarity, the markings which would normally be on the photograph are shown separately on the next page.

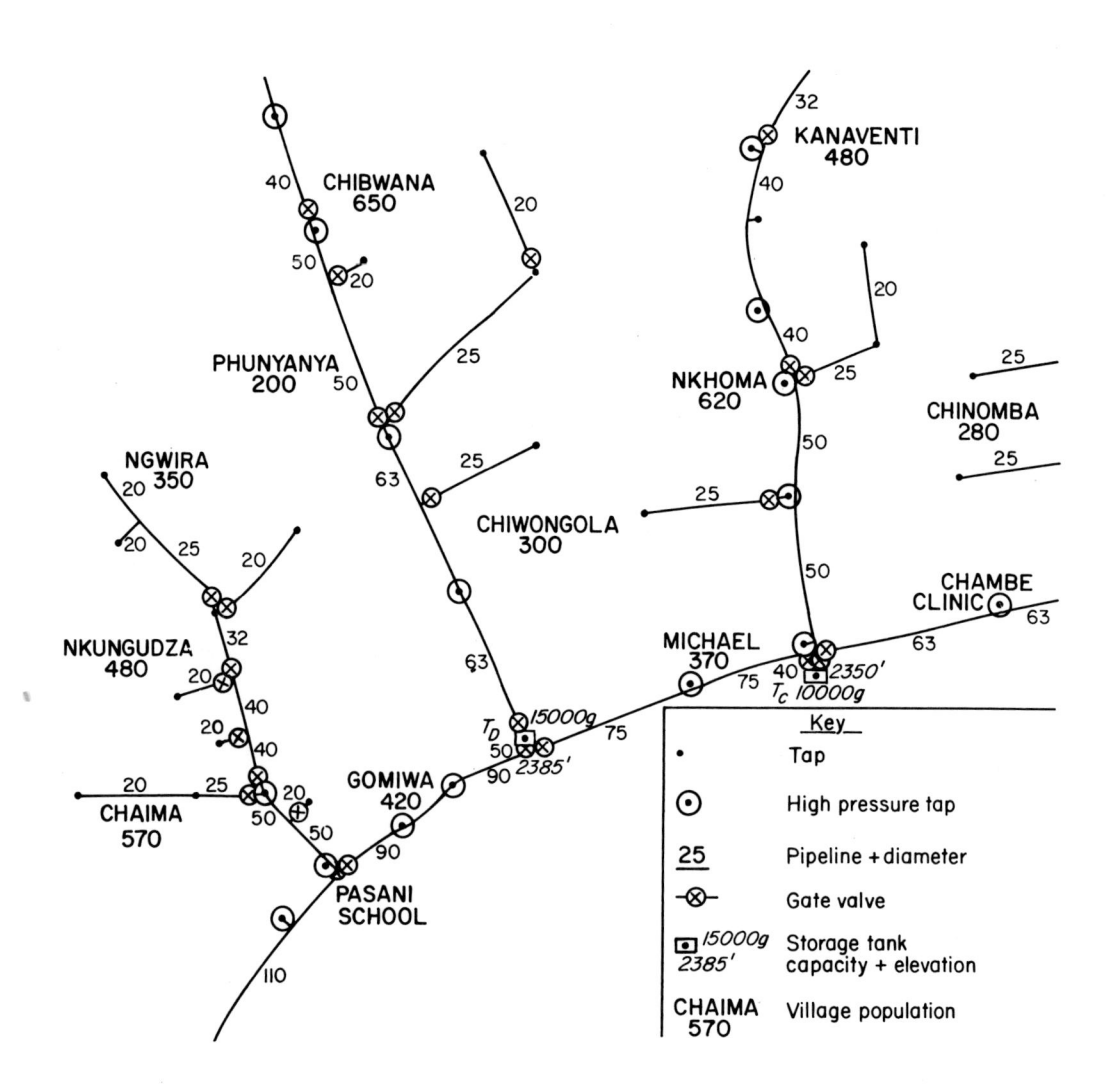

Bibliography

Cairncross, A. M., and Feachem, R. G. (1978). *Small Water Supplies*. Ross Bulletin No. 10. Ross Institute of Tropical Hygiene, London.

Cairncross, A. M., Carruthers, I., Curtis, D., Feachem, R., Bradley, D., and Baldwin, G. (1980). *Evaluation for Village Water Supply Planning*. Wiley, Chichester.

Elmendorf, M., and Buckles, P. (1980). *Sociocultural Aspects of Water Supply and Excreta Disposal*. Appropriate Technology for Water Supply and Sanitation, Volume 5. Transportation, Water and Telecommunications Department, World Bank, Washington.

Feachem, R. G., Burns, E., Cairncross, A., Cronin, A., Cross, P., Curtis, D., Khan, M., Lamb, D., and Southall, H. (1978). *Water, Health and Development* Tri-Med Books, London.

Feachem, R. G., McGarry, M., and Mara, D. (eds) (1977). *Water, Wastes and Health in Hot Climates*. Wiley, Chichester.

Glennie, C. (1982). *A Model for the Development of a Self-Help Water Supply Program*. Technology Advisory Group Working Paper No. 1. World Bank, Washington.

Huisman, L., De Azevedo Netto, J., Sundaresan, B., Lanoix, J., and Hofkes, E. (1981). *Small Community Water Supplies: Technology of Small Water Supply Systems in Developing Countries*. Technical Paper No. 18. WHO International Reference Centre for Community Water Supply and Sanitation, The Hague.

International Association on Water Pollution Research (1979). *Engineering, Science and Medicine in the Prevention of Tropical Water-Related Disease*. Pergamon Press, Oxford.

Jordan, T. (1980). *Handbook of Gravity Flow Systems*. UNICEF, Kathmandu.

Kalbermatten, J., Julius, D., and Gunnerson, C. (1980a). *Technical and Economic Options*. Appropriate Technology for Water Supply and Sanitation, Volume 1. Transportation, Water, and Telecommunications Department, World Bank, Washington.

Kalbermatten, J., Julius, D., and Gunnerson, C. (1980b). *A Planner's Guide*. Appropriate Technology for Water Supply and Sanitation, Volume 2. Transportation, Water, and Telecommunications Department, World Bank, Washington.

Lauria, D., Kolsky, P., and Middleton, R. (1980). *Design of Low-Cost Water Distribution Systems*. Appropriate Technology for Water Supply and Sanitation, Volume 9. Transportation Water, and Telecommunications Department, World Bank, Washington.

Miller, D. (1979). *Self-help and Popular Participation in Rural Water Systems*. OECD, Paris.

Pacey, A. (ed.) (1977). *Water for the Thousand Millions*. Intermediate Technology Development Group, Water Panel. Pergamon Press, Oxford.

Saunders, R. J. and Warford, J. J. (1976). *Village Water Supply: Economics and Policy in the Developing World*. Johns Hopkins University Press, Baltimore.

Spangler, C. (1980). *Low-Cost Water Distribution: A Field Manual*. Appropriate Technology for Water Supply and Sanitation, Volume 12. Transportation, Water, and Telecommunications Department, World Bank, Washington.

UNICEF (1979). *People, Water and Sanitation*. Assignment Children, 45/46. UNICEF, Geneva.

White, A. (1981). *Community Participation in Water and Sanitation: Concepts, Strategies and Methods*. Technical Paper No. 17. WHO International Reference Centre for Community Water Supply, The Hague.

White, G. F., Bradley, D. J., and White, A. U. (1972). *Drawers of Water: Domestic Water Use in East Africa*. University of Chicago Press, Chicago.

Wijk-Sijbesma, C. van (1979a). *Participation and Education in Community Water Supply and Sanitation Programmes: A Literature Review*. Technical Paper No. 12. WHO International Reference Centre for Community Water Supply, The Hague.

Wijk-Sijbesma, C. van (1979b). *Participation and Education in Community Water Supply and Sanitation Programmes: A Selected and Annotated Bibliography*. Bulletin No. 13. WHO International Reference Centre for Community Water Supply, The Hague.

World Bank (1976). *Village Water Supply*. A World Bank Paper. World Bank, Washington.

WHO/World Bank (1978). *Water Supply and Sewerage Sector Study*. Report to the Government of Malawi.

WHO (1979a). *Public Standpost Water Supplies*. Technical Paper No. 13. WHO International Reference Centre for Community Water Supply, The Hague.

WHO (1979b). *Public Standpost Water Supplies: A Design Manual*. Technical Paper No. 14. WHO International Reference Centre for Community Water Supply, The Hague.

Index

Aerial photographs, 82, 142, 146–147
Air valves, 86, 142
Applications for projects, 29, 116
Asbestos cement pipes, 51, 91–92, 107

Backfilling, 52–53
Benefits, 34, 110, 111–114

Cash contributions, 104
Committees, 43–45, 117
Community participation, 37–39, 42–56, 105, 124
Consolidation phase, 19–21, 126–127
Construction of tanks, 58–59, 62
Cost, per capita, 33–34

Decade, 2–3, 123–129
Demand, 22–23, 124–125
Demonstration effect, 19, 21
Demonstration latrines, 119
Design, 79–82, 114, 137–144

Engineers, 76–78, 133–134
Erosion of pipelines, 86, 89, 106
Evaluation, 109–110
Expansion phase, 21, 23, 127–128

Feasibility surveys, 31–33, 116–117
Flow in pipelines, 81–82, 140
Flushing points, 86, 142

HDPE pipe, 94
Health benefit, 112–114
Health education, 114
Health posts, 119

Initiative for projects, 29, 31
Intakes, 83, 105–106

Job descriptions, 133–135

Maintenance, 97–109
Maintenance phase, 128–129
Malawi, background to, 4–9
    origins of programme in, 11–16
Management, of community, 45–56, 124
    of staff, 56–58, 123–124
Maps, 82, 145
Meetings, 42–43
Ministry, choice of, 28
Monitoring, 109
Motivation of staff, 73–75

Personnel policy, 65–66
Pilot phase, 11–19, 124–126
Pipelaying, organization, 51–53
Pipelines, 86, 89, 91–4, 106–107, 140–141
Procurement, 35
Programme management, 3, 25–35, 124–129
Project management, 16–18, 33–35, 38–42
PVC pipe, 53, 92–93, 107–108

Recruitment of staff, 67–68, 75, 76–77

Sanitation, 115–122
Screening tank, 83
Seasonal factors, 39
Sedimentation, 85
Selection criteria, for projects, 33–34
    for staff, 66–67
Self-help, 45–53, 56, 64
Staff, management and training, 56–58, 65–78, 123, 133–135
Standard procedures, 20, 57, 124, 126–127
Standards, technical, 20, 97–98, 127
Steel pipes, galvanized, 83, 86, 94
Stores organization, 64, 104
Supervision, 17, 37, 104, 124
Supervisors, 75–76, 135
Supplies, 35, 143–144
Surveys, feasibility, 31–33

151

Tanks, 59–62, 83–87, 106, 139, 142
Taps, 79–80, 88, 89, 108–109
Technical standards, 20, 97–98, 127
Training, engineers, 77–78
   supervisors, 75–76
   technicians, 68–72

Transport requirements, 62–3

Valves, 89, 143
Village committees, 45
Village Maintenance Worker, 105
Village taps, 79–80, 88, 89, 108–109, 138